Intracellular Traffic and Neurodegenerative Disorders

RESEARCH AND PERSPECTIVES IN ALZHEIMER'S DISEASE

Peter H. St. George-Hyslop • William C. Mobley
Yves Christen
Editors

Intracellular Traffic and Neurodegenerative Disorders

Editors

Dr. Peter H. St. George-Hyslop
Department of Laboratory Medicine
and Pathobiology
University of Toronto
Tranz Neuroscience Bldg.
Toronto ON M5S 3H2
Canada
p.hyslop@utoronto.ca

Dr. William C. Mobley
Department of Neurology
Standford University School of Medicine
Standford CA 94305-5316
USA
ngfv1su@yahoo.com

Dr. Yves Christen
Fondation IPSEN
Pour la Recherche Thérapeutique
65, quai Georges Gorse
92650 Boulogne Billancourt
Cedex - France
yves.christen@ipsen.com

ISSN 0945-6066
ISBN 978-3-540-87940-4 e-ISBN 978-3-540-87941-1

Library of Congress Control Number: 2008936139

Printed on acid-free paper

springer.com

Foreword

Neurodegenerative disorders are common and devastating. Rationally, the most effective treatments will target pathogenetic mechanisms. While alternative approaches, based on alleviating the symptoms of patients with Alzheimer disease, Parkinson disease, Huntington disease, prion disorders or amyotrophic lateral sclerosis, can be expected to reduce suffering, studies of pathogenesis of these age-related disorders will be most important for enabling early diagnosis and the creation of preventative and curative treatments. It is in this context that a recent IPSEN meeting (The 23rd *Colloque Médecine et Recherche*, April 28, 2008) focused on a role for disruption of intracellular trafficking in neurodegenerative disorders. The meeting captured emerging insights into pathogenesis from disrupted trafficking and processing of proteins implicated in age-related degeneration.

Protein folding, trafficking and signaling were the principal topics covered at the meeting. Importantly, the presenters pointed to the importantly intersection of these themes. While the proteolytic processing of APP into its toxic product, the Aβ peptide, is an intensive focus of work in many laboratories, it is only relatively recently that investigators have begun to examine in depth the cellular compartments and trafficking events that mediate APP processing and how derangement of trafficking pathways could impact them. Thus, discoveries by St George-Hyslop and colleagues that SORL1 binds APP, that certain polymorphisms in SORL1 increases the risk of Alzheimer disease and that several of these polymorphisms are predicted to modify SORL1 levels so as to increase Aβ production provided the perspective that malfunction of cellular mechanisms could play a defining role in APP-linked pathology. Willnow built on this theme by defining further the cellular pathways impacted by SORLA, while Seaman linked these observations with proteins of the retromer complex, for which earlier evidence suggested a link to altered APP processing. Contributions by Beyreuther and Kins and by Haass further informed the discussion by providing new insights into the proteins with which APP interacts, including its family members APLP1 and 2, and through studies of g secretase. Gandy reviewed studies showing that APP sorting and metabolism is informed by a number of extracellular signals that act through phosphorylation of APP. Importantly, the participation of the endosomal pathway and early endosomes in particular reinforce the view that trafficking errors at this locus contribute significantly to APP-linked pathology, observations addressed directly by Rajendran and Simons. Sorkin detailed recent advances in understanding protein trafficking and signaling in the endosomal system, studies that must now be extended to APP. But what is it about APP misprocessing that defines key steps in pathogenesis? Most investigators focus squarely on Aβ, but recent findings suggest that a more refined focus on APP will be needed to understand important steps. Indeed, Mobley and colleagues, in studies of mouse models of Down syndrome, show that APP gene dose, and particularly the levels of its C-terminal fragments, may be more directly linked to Alzheimer-like pathogenesis than the level of the Aβ peptide. By what mechanisms would altered trafficking mechanisms influence the cell? An emerging theme, one that links studies of Alzheimer pathogenesis to other neurodegenerative disorders, is that protein misfolding plays a defining role. This was the focus of work reported by Lindquist, in studies of Parkinson and Huntington disease models, and Mandelkow and colleagues in studies of tau mutants. The ability of misfolded proteins to dysregulate cellular processes raises the exciting possibility that protein misfolding errors can be defined and serve as a target of future therapeutics. In the end, it will be essential to explore the events whose compromise is critical to neural cell survival and function. One important lesion may be the axonal transport of trophic messages. Holzbauer makes a compelling case that such messages are markedly compromised in models of amyotrophic lateral sclerosis and Saudou documents dramatic changes in BDNF trafficking in models of Huntington disease. Finally, Mobley reports disruption of NGF transport in models of Down syndrome and Alzheimer disease. That other important retrograde messages must be examined is suggested by Martin and colleagues who document the dynamic processes that link axonal transport with synaptic plasticity.

Though it is difficult to predict the course of future work, the meeting supported the view that misregulation of processing and trafficking events, especially those that occur in the endocytic pathway, will be important for defining and countering the pathogenesis of age-related neurodegenerative disorders.

W. Mobley
P. St George-Hyslop
Y. Christen

Acknowledgements

The editors wish to thank Jacqueline Mervaillie and Sonia Le Cornec for the organization of the meeting and Mary Lynn Gage for the editing of the book.

Contents

Contributors

Allendoerfer Karen L.
Whitehead Institute for Biomedical Research and Howard Hughes Medical Institute, 9 Cambridge Center, Cambridge MA 02142, USA

da Cruz e Silva Edgar
Centro de Biologia Celular, University of Aveiro, Aveiro, Portugal

da Cruz e Silva Odete
Centro de Biologia Celular, University of Aveiro, Aveiro, Portugal

Ehrlich Michelle
Mount Sinai School of Medicine
New York NY 10029

Farrer Lindsay A.
Departments of Medicine (Genetics Program), Neurology, Genetics & Genomics, Epidemiology, and Biostatistics, Boston University Schools of Medicine and Public Health, Boston MA 02118, USA

Fluhrer Regina
Center for Integrated Protein Science Munich and Adolf-Butenandt-Institute, Department of Biochemistry, Laboratory for Neurodegenerative Disease Research, Ludwig-Maximilians-University, 80336 Munich, Germany

Gandy Samuel E.
Mount Sinai School of Medicine
New York NY 10029
samuel.gandy@mssm.edu

Haass Christian
Center for Integrated Protein Science Munich and Adolf-Butenandt-Institute, Department of Biochemistry, Laboratory for Neurodegenerative
Disease Research, Ludwig-Maximilians-University, 80336 Munich, Germany,
chaass@med.uni-muenchen.de

Holzbaur Erika L.F.
Department of Physiology, University of Pennsylvania School of Medicine, D400 Richards Building, 3700 Hamilton Walk, Philadelphia PA 19104, USA, holzbaur@mail.med.upenn.edu

Humbert Sandrine
UMR 146 CNRS, Institut Curie, Bâtiment 110-Centre Universitaire, 91405 Orsay, France

Konzack S.
Max-Planck-Unit for Structural Molecular Biology, c/o DESY, Notkestrasse 85, 22607 Hamburg, Germany

Lai Kwok-On
Department of Psychiatry and Biobehavioral Sciences, Brain Research Institute, UCLA, BSRB 390B, 615 Charles E. Young Dr. S., Los Angeles CA 90095-1737, USA

Lee Joseph H.
The Taub Institute on Alzheimer's Disease and the Aging Brain, The Gertrude H. Sergievsky Center, College of Physicians Surgeons, Department of Epidemiology, Mailman School of Public Health, Columbia University, New York, USA

Lindquist Susan
Whitehead Institute for Biomedical Research and Howard Hughes Medical Institute, 9 Cambridge Center, Cambridge MA 02142, USA, lindquist_admin@wi.mit.edu

Mandelkow Eckhard
Max-Planck-Unit for Structural Molecular Biology, c/o DESY, Notkestrasse 85, 22607 Hamburg, Germany

Mandelkow Eva-Maria
Max-Planck-Unit for Structural Molecular Biology, c/o DESY, Notkestrasse 85, 22607 Hamburg, Germany, mandelkow@mpasmb.desy.de

Martin Kelsey
Department of Psychiatry and Biobehavioral Sciences, Brain Research Institute, Department of Biological Chemistry, Semel Institute for Neuroscience and Human Behavior, UCLA, BSRB 390B, 615 Charles E. Young Dr. S., Los Angeles CA 90095-1737 USA, kcmartin@mednet.ucla.edu

Mayeux Richard
The Taub Institute on Alzheimer's Disease and the Aging Brain, The Gertrude H. Sergievsky Center, College of Physicians Surgeons, Department of Epidemiology, Mailman School of Public Health, Columbia University, New York, USA

Meng Yan
Departments of Medicine (Genetics Program), Neurology, Genetics & Genomics, Epidemiology, and Biostatistics, Boston University Schools of Medicine and Public Health., Boston MA 02118, USA

Mobley William
Department of Neurology, MSLS, P205, Stanford University School of Medicine, 300 Pasteur Drive Stanford CA 94305, USA, ngfv1su@yahoo.com

Rajendran Lawrence
Max Planck Institute of Molecular Cell Biology and Genetics, Pfotenhauerstrasse 108, 01307 Dresden, Germany, rajendra@mpi-cbg.de

Rogaeva Ekaterina
Centre for Research in Neurodegenerative Diseases, Departments of Medicine, Laboratory Medicine and Pathobiology, Medical Biophysics, University of Toronto, and Toronto Western Hospital Research Institute, Toronto, Ontario, Canada

Rohe Michael
Max-Delbrueck-Center for Molecular Medicine, Berlin, Germany

Salehi Ahmad
Stanford University School of Medicine, Dept of Neurology, Stanford CA 94305, USA, ASalehi@stanford.edu

Saudou Frédéric
UMR 146 CNRS, Institut Curie, Bâtiment 110-Centre Universitaire, 91405 Orsay, France, frederic.saudou@curie.u-psud.fr

Schmidt Vanessa
Max-Delbrueck-Center for Molecular Medicine, Berlin, Germany

Seaman Matthew
Department of Clinical Biochemistry, Cambridge Institute for Medical Research, Wellcome Trust & MRC Building, Addenbrookes Hospital, Hills Road, Cambridge CB2 2XY, UK, mnjs100@cam.ac.uk

Simons Kai
Max Planck Institute of Molecular Cell Biology and Genetics, Pfotenhauerstrasse 108, 01307 Dresden, Germany, simons@mpi-cbg.de

Skinner Claire F.
Department of Clinical Biochemistry, Cambridge Institute for Medical Research, Wellcome Trust & MRC Building, Addenbrookes Hospital, Hills Road, Cambridge CB2 2XY, UK

Small Scott
Columbia University College of Physicians and Surgeons,
New York NY 10032, USA

Sorkin Alexander
Department of Pharmacology, University of Colorado at Denver and Health, Sciences Center, Room 6115, Research Complex 1, 12800 East 19th Avenue, Aurora CO 80045, USA, alexander.sorkin@uchsc.edu

St George-Hyslop Peter
Centre for Research in Neurodegenerative Diseases, Departments of Medicine, Laboratory Medicine and Pathobiology, Medical Biophysics, University of Toronto, and Toronto Western Hospital Research Institute, Toronto, Ontario, Canada and Cambridge Institute for Medical Research and Dept of Clinical Neurosciences, University of Cambridge, Wellcome Trust /MRC Building, Addenbrookes Hospital, Hills Road, Cambridge CB2 0XY, UK, p.hyslop@utoronto.ca

Suzuki Toshiharu
Hokkaido University, Sapporo, Japan

Thies E.
Max-Planck-Unit for Structural Molecular Biology, c/o DESY, Notkestrasse 85, 22607 Hamburg, Germany

Wang Dan
Department of Psychiatry and Biobehavioral Sciences, Brain Research Institute, UCLA, BSRB 390B, 615 Charles E. Young Dr. S., Los Angeles CA 90095-1737, USA

Willnow Thomas
Max-Delbrueck-Center for Molecular Medicine, Robert-Roessle-Str. 10, D-13125 Berlin, Germany, willnow@mdc-berlin.de

Wu Chengbiao
Stanford University School of Medicine, Dept of Neurology, CA 94305, Stanford, USA

Amyloid Precursor Protein Sorting and Processing: Transmitters, Hormones, and Protein Phosphorylation Mechanisms*

Sam Gandy(✉), Odete da Cruz e Silva, Edgar da Cruz e Silva, Toshiharu Suzuki, Michelle Ehrlich, and Scott Small

Abstract Since the late 1980's, protein phosphorylation-mediated mechanisms have been recognized as regulators of sorting and processing of the Alzheimer's amyloid precursor (APP). These phospho-state-sensitive steps, in turn, determine the quality and quantity of Aβ generation. Here, we review several recent advances in this field, including new evidence that: (1) the phospho-state of APP threonine-668 does not obviously regulate APP sorting, Aβ generation or Aβ speciation; (2) β-secretase (BACE) recycling is regulated by the phospho-state of the BACE cytoplasmic tail, but without impact on Aβ generation or speciation; (3) contrary to its well-documented acute actions, chronic protein kinase C activation *increases* Aβ generation; and (4) sorting of APP and/or its α- and β-carboxyl-terminal fragments (C83 and C99, respectively) toward the *trans*-Golgi network is under the influence of presenilins and the VPS35/retromer. With the recent discovery of genetic linkage between the risk for Alzheimer's disease (AD) and polymorphisms in *SORL1*, a gene belonging to the sortilin class of trafficking proteins, the membrane protein cell biology of APP has emerged as a central focus for investigators seeking to understand the basis of common forms of AD and thereby uncover new therapeutic opportunities for its treatment and/or prevention.

The phosphorylation states of membrane proteins, such as the Alzheimer's amyloid precursor protein (APP) or β-APP-site cleaving enzyme (BACE), and/or the phosphorylation states of their specific interacting proteins provide for dynamic regulation of signal transduction and protein sorting on a moment-to-moment basis, thereby integrating protein sorting and neurotransmission (Mostov and Cardone 1995; Clague and Urbe 2001; Bonifacino and Traub 2003). A striking example

---

*Reprinted in part from *Neuron* (2006), with permission.

S. Gandy
Mt Sinai School of Medicine, New York NY 10029
E-mail: samuel.gandy@mssm.edu

P. St. George-Hyslop et al. (eds.) *Intracellular Traffic and Neurodegenerative Disorders*, Research and Perspectives in Alzheimer's Disease,

is that of regulated ectodomain shedding of APP (Buxbaum et al. 1990, 1992; Caporaso et al. 1992; Nitsch et al. 1992; Gillespie et al. 1992; Pedrini et al, 2005).

During regulated shedding, first messengers, such as neurotransmitters and hormones (Buxbaum et al. 1992; Nitsch et al. 1992; Jaffe et al. 1994; Xu et al. 1998; Qin et al. 2006), impinge upon neurons and direct APP toward the cell surface and away from the TGN and endocytic pathways (Xu et al. 1995), and hence away from BACE. At the cell surface, APP can be processed by a nonamyloidogenic pathway, known as the α-secretase pathway and defined by the metalloproteinases, ADAM-9, ADAM-10 and ADAM-17 (Buxbaum et al. 1998b; Esler and Wolfe 2001; Allinson et al. 2003; Postina et al. 2004; Kojro and Fahrenholz 2005). ADAM is an acronym derived from "a disintegrin and metalloproteinase."

The molecular mechanism of regulated shedding remains to be fully elucidated but appears to involve phosphorylation of components of the trans-Golgi Network (TGN) vesicle biogenesis machinery (thereby increasing APP delivery to the cell surface; Xu et al. 1995) as well as phosphorylation of protein components of the endocytic system (thereby blocking APP internalization; Chyung and Selkoe 2003; Carey et al. 2005). The phosphorylation states of APP and BACE do not appear to be involved in this process (Gandy et al. 1988; Oishi et al., 1997; da Cruz e Silva et al. 1993; Jacobsen et al. 1994; Pastorino et al. 2002; Ikin et al. 2007). With regard to Aβ generation, this phenomenon is noteworthy because hyperactivation of the α-pathway (e.g., with a combination of simultaneous protein kinase activation and protein phosphatase inhibition) can lead to relatively greater cleavage of APP by α-secretase(s) (Caporaso et al. 1992; Gillespie et al. 1992), thereby reducing or completely abolishing Aβ generation (Buxbaum et al. 1993; Gabuzda et al. 1993; Hung et al. 1993). Interest in this phenomenon has recently been revived with the demonstration that microdialysis techniques can be used to demonstrate and quantify regulated shedding and regulated Aβ generation in the brains of living experimental animals (Cirrito et al. 2005, 2008).

Recent evidence suggests that axonal transport of APP (Lee et al. 2003) and perhaps also prolyl isomerization might be modulated by the state of phosphorylation of the APP cytoplasmic tail at threonine-668 (Pastorino et al. 2006). APP is axonally transported in holoprotein form (Koo et al. 1990; Buxbaum et al. 1998a); hence, the phosphorylation of threonine-668 was proposed to serve as a "tag," targeting phospho-forms of APP for delivery to the nerve terminal (Lee et al. 2003). However, recent evidence calls into question the proposal that the phosphorylation state of threonine-668 plays a major physiological role in APP localization or Aβ generation, since threonine-to-alanine-668 knock-in mice show normal levels and subcellular distributions of APP and its metabolites, including Aβ (Sano et al. 2006). There is compelling evidence, however, that, once at the nerve terminal, APP is processed, generating Aβ locally at the terminal and releasing Aβ at, near or into the synapse (Kamenetz et al. 2003).

The cytoplasmic tail of BACE also undergoes reversible phosphorylation, and that event appears to specify its recycling (von Arnim et al. 2004; He et al. 2005). In cell lines, the dephospho- and phospho-forms of BACE appear to perform with similar efficiencies in generating Aβ40 and Aβ42 (Pastorino et al. 2002), but this finding has not been evaluated in primary neuronal cultures. This failure of Aβ generation to be regulated by BACE recycling is somewhat unexpected since, as reviewed above, most Aβ is believed to arise from the endocytic pathway. Hence, one would expect that increasing BACE concentration in the endocytic pathway would increase generation of Aβ. One explanation for this unexpected result is that the substrate may be limiting in post-TGN compartments, and therefore increased levels of BACE are unable to raise Aβ generation. This notion agrees with the proposal mentioned above that regulated shedding acts at the TGN to divert APP molecules toward the plasma membrane as a means of lower generation of Aβ, at least in part because a limited pool of APP is transported out of the TGN (Buxbaum et al. 1993; Skovronsky et al. 2000). Indeed, in some neuron-like cell types, over 80% of the newly synthesized moles of APP are degraded without generating obvious, discrete metabolic fragments (Caporaso et al. 1992).

Clathrin-independent endocytosis of transmembrane proteins is regulated by protein phosphorylation (Robertson et al. 2006). Further, two components of the endocytosis machinery, dynamin and amphiphysin, control clathrin-mediated endocytosis in a fashion that is sensitive to their direct phosphorylation by the protein kinase cdk5 (Tomizawa et al. 2003; Nguyen and Bibb 2003). Retromer function is regulated by a separate complex of molecules known as "complex II" (Burda et al. 2002). Complex II includes several catalytic functions that direct retromer action. The phosphoinositide kinase VPS34 binds the protein kinase VPS15, and then, secondarily, VPS30 and VPS38 are recruited and the four molecules comprise the complete complex II (Burda et al. 2002). Thus, complex II action is modulated not only by protein phosphorylation but also by lipid phosphorylation (Stack et al. 1995). Some investigators have proposed that the PI3-kinase component of complex II directs synthesis of a specific pool of endosomal PI3, which, in turn, activates or stimulates assembly of the retromer complex, thereby ensuring efficient endosome-to-Golgi retrograde transport (Stack et al. 1995). These regulatory mechanisms may have implications for Aβ generation, but such a connection, if one exists, remains to be elucidated.

Presenilins may also modulate protein trafficking and sorting. Soon after the discovery of presenilins, gene-targeting experiments were performed in mice to investigate the essential bioactivities of these complex, polytopic, molecules, especially presenilin 1 (PS1; Wong et al. 1997; Naruse et al. 1998). In cells from PS1-deficient mice, delivery of multiple type-I proteins to the cell surface was observed to be disturbed; APP and the p75 neurotrophin receptor were among those missorted proteins (Naruse et al. 1998). This work was somewhat overshadowed, however, when cells from PS1-deficient mice were demonstrated to be incapable of generating Aβ (DeStrooper et al. 1998). This observation placed APP and PS1 on a common metabolic pathway for the first time and was rapidly followed by demonstration that PS1 did, indeed, contain the catalytic site of γ-secretase, as established by cross-linking of γ-secretase inhibitors to PS1 (Li et al. 2000a, b).

The unusual intramembranous localization of two aspartate residues led to the postulation that these amino acids were forming the active site of an aspartyl proteinase (Wolfe et al. 1999). This explanation dovetailed with the apparent fact that

APP C-terminal fragments were cleaved by regulated intramembranous proteolysis (RIP), and when the aspartates were mutated to alanines, γ-secretase activity was abolished (Wolfe et al. 1999). RIP was, at the time, a relatively recently recognized phenomenon, and conventional wisdom up to that point had held that the hydrophobicity of membranes would preclude the entry of water into the lipid bilayer to enable hydrolysis of peptide bonds. Even to this day, the mechanism that provides the capability for surmounting that energy barrier is poorly understood. The popular formulation at that point was that PS1 was a proteinase, and the notion that PS1 was a trafficking factor was underemphasized. The possibility was also raised that aberrant trafficking in PS1 deficient cells was perhaps due to the inability of some unidentified PS1 substrate trafficking factor to function properly in its uncleaved state, since its cognate protease (PS1) was absent.

Beginning in the last few years, however, experiments in cultured cells and cell-free assays have begun to yield consistent, compelling evidence that PS1 bears a trafficking function in addition to its catalytic function, or, alternatively, as mentioned above, that trafficking proteins were important substrates for cleavage by PS1 so that, when PS1 was deficient, post-TGN trafficking of membrane protein cargo became abnormal (Kaether et al. 2002; Wang et al. 2004; Wood et al. 2005; Rechards et al. 2006).

Most PS1-deficient mice and cells are highly compromised and resemble Notch-deficient mice and cells (Wong et al. 1997). This finding is not entirely unexpected since Notch is a substrate for cleavage by γ-secretase, as are another several dozen type-I transmembrane proteins, including cadherin, erb-b4, and the p75 NGF receptor (DeStrooper et al. 1999; Struhl and Greenwald 1999; for review, see Fortini 2002). Therefore, PS1-deficiency can lead to dysfunction of a host of proteins whose physiological function requires cleavage by RIP to release their cytoplasmic domains. In many examples, the cytoplasmic domain released by γ-secretase appears to diffuse rapidly to the nucleus, where these intracellular domains (ICDs), such as Notch intracellular domain (NICD), modulate gene transcription (Cupers et al. 2001; Fortini 2002; Cao and Sudhof 2001).

PS1-mediated trafficking appears to localize to post-TGN steps of trafficking of type I transmembrane proteins (Annaert et al. 1999; Kaether et al. 2002; Wang et al. 2004; Wood et al. 2005; Wang et al. 2006; Zhang et al. 2006; Cai et al. 2003, 2006a, b; Gandy et al. 2007). This role for PS1 in regulation of APP trafficking has been implicated in both cell culture and cell-free in vitro reconstitution studies (Annaert et al. 1999; Kaether et al. 2002; Wang et al. 2004; Wood et al. 2005; Wang et al. 2006; Zhang et al. 2006; Cai et al. 2003, 2006a, b; Gandy et al. 2007). Pathogenic PS1 mutations retard egress of APP from the TGN by a mechanism that appears to involve phospholipase D (Cai et al. 2006a, b), a known TGN budding modulator (Kahn et al. 1993). It is clear that the mutations that have been tested so far increase the residence time at the TGN while also increasing the Aβ42/40 ratio (Kahn et al. 1993). Recent data suggest that TGN retention per se can increase generation of Aβ 42/40 in cerebral neurons in vivo, indicating that abnormal post-TGN trafficking of APP might be sufficient to initiate Aβ accumulation (Gandy et al. 2007).

The pathogenic PS1 defect can be corrected in cell culture and in cell-free systems following supplementation of the budding factor phospholipase D (PLD; Cai et al. 2003, 2006a, b). The molecular details of how PS1 and PLD are connected remain obscure; however, as cargos other than APP are found to be missorted, including, e.g., tyrosinase (Wang et al. 2006), the notion that PS1 has a protein trafficking function has become more widely appreciated and accepted. Now, the challenge is to identify at the molecular level those factors that selectively favor cleavage at the Aβ42–43 scissile bond.

PS1 has also been implicated in trafficking of APP and perhaps its carboxyl terminal fragments out of the endosome (Zhang et al. 2006). Thus, PS1 dysfunction could also result in retention of APP and CTFs within the endocytic compartment, which, in turn, would favor Aβ generation. Thus, accumulating evidence implicates PS1 in the regulation of APP trafficking. The possibility exists that the local environment within the TGN or the endocytic system contributes to misalignment of mutant PS1 and APP carboxyl terminal fragments, thereby favoring generation of Aβ42. Such a mechanism has been implicated in other diseases (e.g., cystic fibrosis) that are also caused by missense mutations in polytopic proteins (Gentzsch et al. 2004).

In conclusion, elucidating the mechanisms that sort APP and the secretases through the TGN, cell surface, and endosome has significantly expanded the understanding of Alzheimer's disease cell biology. More importantly, isolating specific defects in protein sorting opens up unexplored therapeutic avenues that, optimistically, may accelerate the development of effective treatments for this devastating and intractable disease.

Acknowledgements The authors acknowledge the support of the Cure Alzheimer's Fund (S.G.), the EU VI Framework Program cNEUPRO (E.C.S., O.C.S), the FCT-REEQ/1025/BIO/2004 award (E.C.S., O.C.S.), the McKnight Foundation (S.S.), the McDonnell Foundation (S.S.), and the NIH, including P50 AG08702 (S.S.), R01 AG025161 (S.S.), R01 AG023611 (S.G.), R01 NS41017 (S.G.), P01 AG10491 (S.G.), and the P50 AG005138 Mount Sinai Alzheimer's Disease Research Center (to Mary Sano). We also thank Enid Castro for administrative support.

References

Allinson TM, Parkin ET, Turner AJ, Hooper NM (2003) ADAMs family members as amyloid precursor protein alpha-secretases. J Neurosci Res 74: 342–352.

Annaert WG, Levesque L, Craessaerts K, Dierinck I, Snellings G, Westaway D, George-Hyslop PS, Cordell B, Fraser P, De Strooper B (1999) Presenilin 1 controls gamma-secretase processing of amyloid precursor protein in pre-golgi compartments of hippocampal neurons. J Cell Biol 147: 277–924.

Bonifacino JS, Traub LM (2003) Signals for sorting of transmembrane proteins to endosomes and lysosomes. Annu Rev Biochem 72: 395–447.

Burda P, Padilla SM, Sarkar S, Emr SD (2002) Retromer function in endosome-to-Golgi retrograde transport is regulated by the yeast Vps34 PtdIns 3-kinase. J Cell Sci 115: 3889–3900.

Buxbaum JD, Gandy SE, Cicchetti P, Ehrlich ME, Czernik AJ, Fracasso RP, Ramabhadran TV, Unterbeck AJ, Greengard P (1990) Processing of Alzheimer beta/A4 amyloid precursor protein: modulation by agents that regulate protein phosphorylation. Proc Natl Acad Sci USA 87: 6003–6006.

Buxbaum JD, Oishi M, Chen HI, Pinkas-Kramarski R, Jaffe EA, Gandy SE, Greengard P (1992) Cholinergic agonists and interleukin 1 regulate processing and secretion of the Alzheimer beta/A4 amyloid protein precursor. Proc Natl Acad Sci USA 89: 10075–10078.

Buxbaum JD, Koo EH, Greengard P (1993) Protein phosphorylation inhibits production of Alzheimer amyloid beta/A4 peptide. Proc Natl Acad Sci USA 90: 9195–9198.

Buxbaum JD, Thinakaran G, Koliatsos V, O'Callahan J, Slunt HH, Price DL, Sisodia SS (1998a) Alzheimer amyloid protein precursor in the rat hippocampus: transport and processing through the perforant path. J Neurosci 18: 9629–9637.

Buxbaum JD, Liu KN, Luo Y, Slack JL, Stocking KL, Peschon JJ, Johnson RS, Castner BJ, Cerretti DP, Black RA (1998b) Evidence that tumor necrosis factor alpha converting enzyme is involved in regulated alpha-secretase cleavage of the Alzheimer amyloid protein precursor. J Biol Chem 273: 27765–27767.

Cai D, Leem JY, Greenfield JP, Wang P, Kim BS, Wang R, Lopes KO, Kim SH, Zheng H, Greengard P, Sisodia SS, Thinakaran G, Xu H (2003) Presenilin-1 regulates intracellular trafficking and cell surface delivery of beta-amyloid precursor protein. J Biol Chem 278: 3446–3454.

Cai D, Zhong M, Wang, R, Netzer WJ, Shields D, Zheng H, Sisodia SS, Foster DA, Gorelick FS, Xu H, Greengard P (2006a) Phospholipase D1 corrects impaired betaAPP trafficking and neurite outgrowth in familial Alzheimer's disease-linked presenilin-1 mutant neurons. Proc Natl Acad Sci USA 103: 1936–1940.

Cai D, Netzer WJ, Zhong M, Lin Y, Du G, Frohman M, Foster DA, Sisodia SS, Xu H, Gorelick FS, Greengard P (2006b) Presenilin-1 uses phospholipase D1 as a negative regulator of beta-amyloid formation. Proc Natl Acad Sci USA 103: 1941–1946.

Cao X, Sudhof TC (2001) A transcriptionally [correction of transcriptively] active complex of APP with Fe65 and histone acetyltransferase Tip60. Science 293: 115–120.

Caporaso GL, Gandy SE, Buxbaum JD, Ramabhadran TV, Greengard P (1992) Protein phosphorylation regulates secretion of Alzheimer beta/A4 amyloid precursor protein. Proc Natl Acad Sci USA 89: 3055–3059.

Carey RM, Balcz BA, Lopez-Coviella I, Slack BE (2005) Inhibition of dynamin-dependent endocytosis increases shedding of the amyloid precursor protein ectodomain and reduces generation of amyloid beta protein. BMC Cell Biol 11: 30.

Chyung JH, Selkoe DJ (2003) Inhibition of receptor-mediated endocytosis demonstrates generation of amyloid beta-protein at the cell surface. J Biol Chem 278: 51035–51043.

Cirrito JR, Yamada KA, Finn MB, Sloviter RS, Bales KR, May PC, Schoepp DD, Paul SM, Mennerick S, Holtzman DM. (2005) Synaptic activity regulates interstitial fluid amyloid-beta levels in vivo. Neuron 48(6): 913–22.

Cirrito JR, Kang JE, Lee J, Stewart FR, Verges DK, Silverio LM, Bu G, Mennerick S, Holtzman DM (2008) Endocytosis is required for synaptic activity-dependent release of amyloid-beta in vivo. Neuron 58: 42–51.

Clague MJ, Urbe S (2001) The interface of receptor trafficking and signalling. J Cell Sci 114: 3075–3081.

Cupers P, Orlans I, Craessaerts K, Annaert W, De Strooper B (2001) The amyloid precursor protein (APP)-cytoplasmic fragment generated by gamma-secretase is rapidly degraded but distributes partially in a nuclear fraction of neurones in culture. J Neurochem 78: 1168–1178.

da Cruz e Silva OA, Iverfeldt K, Oltersdorf T, Sinha S, Lieberburg I, Ramabhadran TV, Suzuki T, Sisodia SS, Gandy S, Greengard P (1993) Regulated cleavage of Alzheimer beta-amyloid precursor protein in the absence of the cytoplasmic tail. Neuroscience 57: 873–877.

De Strooper B, Saftig P, Craessaerts K, Vanderstichele H, Guhde G, Annaert W, Von Figura K, Van Leuven F (1998) Deficiency of presenilin-1 inhibits the normal cleavage of amyloid precursor protein. Nature 391: 387–390.

De Strooper B, Annaert W, Cupers P, Saftig P, Craessaerts K, Mumm JS, Schroeter EH, Schrijvers V, Wolfe MS, Ray WJ, Goate A, Kopan R (1999) A presenilin-1-dependent gamma-secretase-like protease mediates release of Notch intracellular domain. Nature 398: 518–522.

Esler WP, Wolfe MS (2001) A portrait of Alzheimer secretases–new features and familiar faces. Science 293: 1449–1454.

Fortini ME (2002) Gamma-secretase-mediated proteolysis in cell-surface-receptor signalling. Nature Rev Mol Cell Biol 3: 673–884.

Gabuzda D, Busciglio J, Yankner BA (1993) Inhibition of beta-amyloid production by activation of protein kinase C. J Neurochem 61: 2326–2329.

Gandy S, Czernik AJ, Greengard P (1988) Phosphorylation of Alzheimer disease amyloid precursor peptide by protein kinase C and Ca2+/calmodulin-dependent protein kinase II. Proc Natl Acad Sci USA 85: 6218–6221.

Gandy S, Zhang YW, Ikin A, Schmidt SD, Bogush A, Levy E, Sheffield R, Nixon RA, Liao FF, Mathews PM, Xu H, Ehrlich ME (2007) Alzheimer's presenilin 1 modulates sorting of APP and its carboxyl-terminal fragments in cerebral neurons in vivo. J Neurochem 102:619–26.

Gentzsch M, Chang XB, Cui L, Wu Y, Ozols VV, Choudhury A, Pagano RE, Riordan JR (2004) Endocytic trafficking routes of wild type and DeltaF508 cystic fibrosis transmembrane conductance regulator. Mol Biol Cell 15: 2684–2696.

Gillespie SL, Golde TE, Younkin SG (1992) Secretory processing of the Alzheimer amyloid beta/A4 protein precursor is increased by protein phosphorylation. Biochem Biophys Res Commun 187: 1285–1290.

He X, Li F, Chang WP, Tang J (2005) GGA proteins mediate the recycling pathway of memapsin 2 (BACE). J Biol Chem 280: 11696–11703.

Hung AY, Haass C, Nitsch RM, Qiu WQ, Citron M, Wurtman RJ, Growdon JH, Selkoe DJ (1993) Activation of protein kinase C inhibits cellular production of the amyloid beta-protein. J Biol Chem 268: 22959–22962.

Ikin AF, Causevic M, Pedrini S, Benson LS, Buxbaum JD, Suzuki T, Lovestone S, Higashiyama S, Mustelin T, Burgoyne RD, Gandy S (2007) Evidence against roles for phorbol binding protein Munc13-1, ADAM adaptor Eve-1, or vesicle trafficking phosphoproteins Munc18 or NSF as phospho-state-sensitive modulators of phorbol/PKC-activated Alzheimer APP ectodomain shedding. Mol Neurodegener 2: 23.

Jacobsen JS, Spruyt MA, Brown AM, Sahasrabudhe SR, Blume AJ, Vitek MP., Muenkel HA, Sonnenberg-Reines J (1994) The release of Alzheimer's disease beta amyloid peptide is reduced by phorbol treatment. J Biol Chem 269: 8376–8382.

Jaffe AB, Toran-Allerand CD, Greengard P, Gandy SE (1994) Estrogen regulates metabolism of Alzheimer amyloid beta precursor protein. J Biol Chem 269: 13065–13068.

Kaether C, Lammich S, Edbauer D, Ertl M, Rietdorf J, Capell A, Steiner H, Haass C (2002) Presenilin-1 affects trafficking and processing of betaAPP and is targeted in a complex with nicastrin to the plasma membrane. J Cell Biol 158: 551–561.

Kahn R.A, Yucel JK, Malhotra V (1993) ARF signaling: a potential role for phospholipase D in membrane traffic. Cell 75: 1045–1048.

Kamenetz F, Tomita T, Hsieh H, Seabrook G, Borchelt D, Iwatsubo T, Sisodia S, Malinow R (2003) APP processing and synaptic function. Neuron 37: 925–937.

Kojro E, Fahrenholz F (2005) The non-amyloidogenic pathway: structure and function of alpha-secretases. Subcell Biochem 38: 105–127.

Koo EH. Sisodia SS, Archer DR, Martin LJ, Weidemann A, Beyreuther K, Fischer, P, Masters CL, Price DL (1990) Precursor of amyloid protein in Alzheimer disease undergoes fast anterograde axonal transport. Proc Natl Acad Sci USA 87: 1561–1565.

Lee MS, Kao SC, Lemere CA, Xia W, Tseng HC, Zhou Y, Neve R, Ahlijanian MK, Tsai LH (2003) APP processing is regulated by cytoplasmic phosphorylation. J Cell Biol 163: 83–95.

Li YM, Lai MT, Xu M, Huang Q, DiMuzio-Mower J, Sardana MK, Shi XP, Yin KC, Shafer JA, Gardell SJ (2000a) Presenilin 1 is linked with gamma-secretase activity in the detergent solubilized state. Proc Natl Acad Sci USA 97: 6138–6143.

Li YM, Xu M, Lai MT, Huang Q, Castro JL, DiMuzio-Mower J, Harrison T, Lellis C, Nadin A, Neduvelil JG, Register RB, Sardana MK, Shearman MS, Smith AL, Shi XP, Yin KC, Shafer JA, Gardell SJ (2000b) Photoactivated gamma-secretase inhibitors directed to the active site covalently label presenilin 1. Nature 405: 689–694.

Mostov KE, Cardone MH (1995) Regulation of protein traffic in polarized epithelial cells. Bioessays 17: 129–138.

Naruse S, Thinakaran G, Luo JJ, Kusiak JW, Tomita T, Iwatsubo T, Qian X, Ginty DD, Price DL, Borchelt DR, Wong PC, Sisodia SS (1998) Effects of PS1 deficiency on membrane protein trafficking in neurons. Neuron 21: 1213–1221.

Nguyen C, Bibb JA (2003) Cdk5 and the mystery of synaptic vesicle endocytosis. J Cell Biol 163: 697–699.

Nitsch RM, Slack BE, Wurtman RJ, Growdon JH (1992) Release of Alzheimer amyloid precursor derivatives stimulated by activation of muscarinic acetylcholine receptors. Science 258: 304–307.

Oishi M, Nairn AC, Czernik AJ, Lim GS, Isohara T, Gandy SE, Greengard P, Suzuki T (1997) The cytoplasmic domain of Alzheimer's amyloid precursor protein is phosphorylated at Thr654, Ser655, and Thr668 in adult rat brain and cultured cells. Mol Med 3: 111–123.

Pastorino L, Ikin AF, Nairn AC, Pursnani A, Buxbaum JD (2002) The carboxyl-terminus of BACE contains a sorting signal that regulates BACE trafficking but not the formation of total A(beta). Mol Cell Neurosci 19: 175–185.

Pastorino L, Sun A, Lu PJ, Zhou XZ, Balastik M, Finn G, Wulf G, Lim J, Li SH, Li X, Xia W, Nicholson LK, Lu KP (2006) The prolyl isomerase Pin1 regulates amyloid precursor protein processing and amyloid-beta production. Nature 440: 528–534.

Pedrini S, Carter TL, Prendergast G, Petanceska S, Ehrlich ME, Gandy S (2005) Modulation of statin-activated shedding of Alzheimer APP ectodomain by ROCK. PLoS Med 2: e18.

Postina R, Schroeder A, Dewachter I, Bohl J, Schmitt U, Kojro E, Prinzen C, Endres K, Hiemke C, Blessing M, Flamez P, Dequenne A, Godaux E, van Leuven F, Fahrenholz F (2004) A disintegrin-metalloproteinase prevents amyloid plaque formation and hippocampal defects in an Alzheimer disease mouse model. J Clin Invest 113: 1456–1464.

Qin W, Yang T, Ho L, Zhao Z, Wang, J, Chen L, Thiyagarajan M, Macgrogan D, Rodgers JT, Puigserver P, Sadoshima J, Deng HH, Pedrini S, Gandy S, Sauve A, Pasinetti GM (2006) Neuronal SIRT1 activation as a novel mechanism underlying the prevention of Alzheimer's disease amyloid neuropathology by calorie restriction. J Biol Chem 281: 21745–21754.

Réchards M, Xia W, Oorschot V, van Dijk S, Annaert W, Selkoe DJ, Klumperman J (2006) Presenilin-1-mediated retention of APP derivatives in early biosynthetic compartments. Traffic 7(3): 354–364.

Robertson SE, Setty SR, Sitaram A, Marks MS, Lewis RE, Chou MM (2006) Extracellular signal-regulated kinase regulates clathrin-independent endosomal trafficking. Mol Biol Cell 17: 645–657.

Sano Y, Nakaya T, Pedrini S, Takeda S, Iijima-Ando K, Iijima K, Mathews PM, Itohara S, Gandy S, Suzuki T (2006) Physiological mouse brain Abeta levels are not related to the phosphorylation state of threonine-668 of Alzheimer's APP. PLoS ONE 1: e51.

Skovronsky DM, Moore DB, Milla ME, Doms RW, Lee VM (2000b) Protein kinase C-dependent alpha-secretase competes with beta-secretase for cleavage of amyloid-beta precursor protein in the trans-golgi network. J Biol Chem 275: 2568–2575.

Stack JH, Horazdovsky B, Emr SD (1995) Receptor-mediated protein sorting to the vacuole in yeast: roles for a protein kinase, a lipid kinase and GTP-binding proteins. Annu Rev Cell Dev Biol 11: 1–33.

Struhl G, Greenwald I (1999) Presenilin is required for activity and nuclear access of Notch in Drosophila. Nature 398: 522–525.

Tomizawa K, Sunada S, Lu YF, Oda Y, Kinuta M, Ohshima T, Saito T, Wei FY, Matsushita M, Li ST, Tsutsui K, Hisanaga S, Mikoshiba K, Takei K, Matsui H (2003) Cophosphorylation of amphiphysin I and dynamin I by Cdk5 regulates clathrin-mediated endocytosis of synaptic vesicles. J Cell Biol 163: 813–824.

von Arnim CA, Tangredi MM, Peltan ID, Lee BM, Irizarry MC, Kinoshita A, Hyman BT (2004) Demonstration of BACE (beta-secretase) phosphorylation and its interaction with GGA1 in cells by fluorescence-lifetime imaging microscopy. J Cell Sci 117: 5437–5445.

Wang H, Luo, WJ, Zhang YW, Li YM, Thinakaran G, Greengard P, Xu H (2004) Presenilins and gamma-secretase inhibitors affect intracellular trafficking and cell surface localization of the gamma-secretase complex components. J Biol Chem 279: 40560–40566.

Wang R, Tang P, Wang P, Boissy RE, Zheng H (2006) Regulation of tyrosinase trafficking and processing by presenilins: partial loss of function by familial Alzheimer's disease mutation. Proc Natl Acad Sci USA 103: 353–358.

Wolfe MS, Xia W, Ostaszewski BL, Diehl TS, Kimberly WT, Selkoe DJ (1999) Two transmembrane aspartates in presenilin-1 required for presenilin endoproteolysis and gamma-secretase activity. Nature. 398: 513–517.

Wood DR, Nye JS, Lamb NJ, Fernandez A, Kitzmann M (2005) Intracellular retention of caveolin 1 in presenilin-deficient cells. J Biol Chem 280: 6663–6668.

Wong PC, Zheng, H, Chen H, Becher MW, Sirinathsinghji DJ, Trumbauer ME, Chen HY, Price DL, Van der Ploeg LH, Sisodia SS (1997) Presenilin 1 is required for Notch1 and DII1 expression in the paraxial mesoderm. Nature 387: 288–292.

Xu H, Greengard P, Gandy S (1995) Regulated formation of Golgi secretory vesicles containing Alzheimer beta-amyloid precursor protein. J Biol Chem 270: 23243–23245.

Xu H, Gouras GK, Greenfield JP, Vincent B, Naslund J, Mazzarelli L, Fried G, Jovanovic JN, Seeger M, Relkin NR, Liao F, Checler F, Buxbaum JD, Chait BT, Thinakaran G, Sisodia SS, Wang R, Greengard P, Gandy S (1998) Estrogen reduces neuronal generation of Alzheimer beta-amyloid peptides. Nature Med 4: 447–451.

Zhang M, Haapasalo A, Kim DY, Ingano LA, Pettingell WH, Kovacs DM (2006) Presenilin/gamma-secretase activity regulates protein clearance from the endocytic recycling compartment. FASEB J 20: 1176–1178.

Intramembrane Proteolysis by γ-Secretase and Signal Peptide Peptidases

Regina Fluhrer and Christian Haass(✉)

Abstract The amyloid cascade hypothesis describes a series of cumulative events that are initiated by amyloid β-peptide and finally lead to synapse and neuron loss. Obviously, the proteases involved in amyloid β-peptide generation are targets for therapeutic treatment strategies. For the development of a safe therapeutic intervention, however, we must understand the precise physiological functions and the cellular mechanisms involved in substrate recognition, selection and cleavage. Moreover, homologous proteases, whose physiological function could be affected by inhibitors, need to be discovered and assays must be developed to help determine the cross-reactive potential of such inhibitors. Here we will focus on the intramembrane cleavage of the β-amyloid precursor protein, which is performed by the γ-secretase complex. In parallel, the cellular and biochemical properties of other proteases belonging to the same family of GxGD-type aspartyl proteases, the signal peptide peptidase and their homologues, will be described. We present a common, multiple intramembrane cleavage mechanism performed by these proteases and evidence that Alzheimer's disease-associated mutations lead to a partial loss of intramembrane proteolysis.

1 Introduction

Alzheimer's disease (AD) is the most prevalent neurodegenerative disorder worldwide (Hardy and Selkoe 2002). The major pathological hallmarks of the disease are senile plaques, composed of amyloid β-peptide (Aβ; Hardy and Selkoe 2002). Aβ is generated from the β-amyloid precursor protein (βAPP) by two sequential endoproteolytic steps. While the first cleavage event, which is mediated by

C. Haass
Center for Integrated Protein Science Munich and Adolf-Butenandt-Institute
Department of Biochemistry, Laboratory for Neurodegenerative Disease Research
Ludwig-Maximilians-University, 80336 Munich, Germany

P. St. George-Hyslop et al. (eds.) *Intracellular Traffic and Neurodegenerative Disorders*,
Research and Perspectives in Alzheimer's Disease,

β-secretase, occurs in the hydrophilic environment of either the extracellular space or the lumen of endosomal/lysosomal/Golgi vesicles, the second cleavage, mediated by γ-secretase, occurs within the hydrophobic environment of cellular membranes. Intramembrane cleavage has been thought to be impossible for quite some time, since it was believed that water molecules, which are absolutely required for proteolysis, are not abundant enough within the hydrophobic bilayer of the membrane. Nonetheless, over the past few years, a number of enzymes have been discovered that share the ability to cleave the transmembrane domain (TMD) of integral membrane proteins (Wolfe and Kopan 2004). These intramembrane cleaving proteases (ICLIPs) are classified according to the amino acid that is localized and required within their catalytically active center. So far representatives of three protease classes have been identified: the site-2 (S2P) metalloprotease (Brown and Goldstein 1999), the GxGD-type aspartyl proteases (Haass and Steiner 2002) and the rhomboid serine proteases (Lemberg and Freeman 2007) (Fig. 1).

ICLIP turned out to be an important part of a novel cellular pathway termed regulated intramembrane proteolysis (RIP). RIP describes the sequential processing of an increasing number of single-pass transmembrane proteins, which as a first step

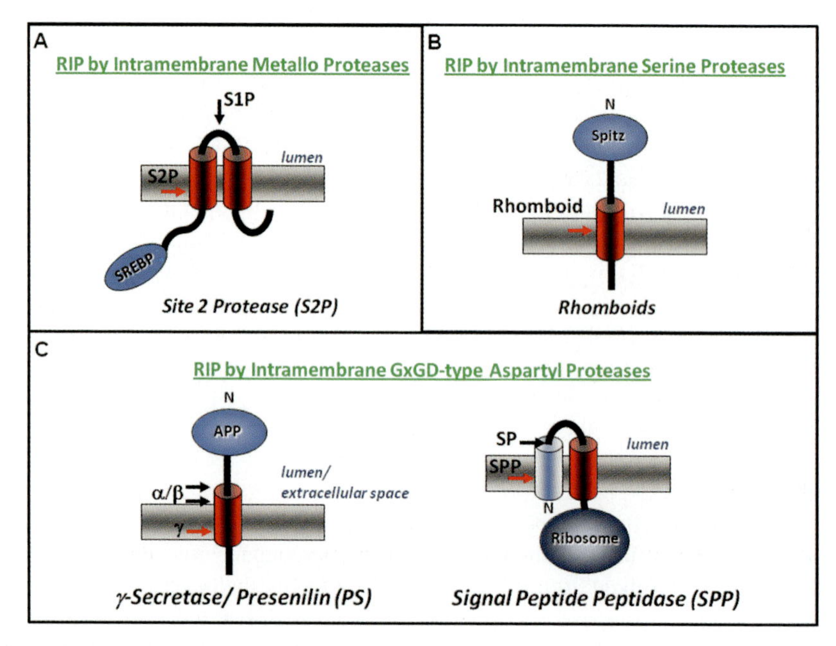

Fig. 1 Models showing regulated intramembrane proteolysis (RIP) by the different classes of intramembrane cleaving proteases. The initial shedding event is marked by a black arrow; the intramembrane cleavage is illustrated by a red arrow. **(A)** RIP of SREBP involving the intramembrane cleaving metallo protease S2P. **(B)** RIP of the *Drosophila melanogaster* protein Spitz involving Rhomboid, an intramembrane cleaving serine protease. **(C)** RIP of βAPP and signal peptides involving γ-secretase and SPP, respectively. γ-Secretase and SPP are representatives of GxGD-type intramembrane cleaving aspartyl proteases

undergo a shedding event, removing large parts of their ectodomain. The remaining membrane-bound stub is subsequently cleaved by an ICLIP within its hydrophobic TMD, releasing small peptides to the extracellular space as well as to the cytosol. Cytosolic peptides, the intracellular domains (ICDs), are in some cases translocated to the nucleus and can be involved in nuclear signaling and transcriptional regulation (Haass 2004; Wolfe and Kopan 2004).

All currently known ICLIPs are polytopic proteins, with their active center most likely embedded within certain TMDs. Apparently this enables these proteases to form water-penetrated cavities, allowing proteolysis within the lipid bilayer of cellular membranes (Feng et al. 2007; Lazarov et al. 2006; Lemberg and Freeman 2007; Steiner et al. 2006).

S2P is required for the regulation of cholesterol and fatty acid biosynthesis via the liberation of the membrane-bound transcription factor, sterol regulatory element binding protein (SREBP), by intramembrane proteolysis. In addition, S2P is involved in intramembrane processing of ATF6, a protein required for chaperone expression during unfolded protein response. Prior to intramembrane cleavage, both substrates are first shed by a luminal cleavage via site-1-protease (S1P; Rawson et al. 1997; Ye et al. 2000; Fig. 1).

A member of the rapidly growing family of rhomboid proteases was first identified in *Drosophila melanogaster* and was shown to be the primary regulator of epidermal growth factor (EGF) receptor signaling via the processing of Spitz, Karen and Gurken (Lemberg and Freeman 2007; Fig. 1). Besides their function in EGF receptor signaling rhomboids are also involved in many other cellular pathways, including apoptosis, generation of a peptidic quorum sensing signal in procaryots, invasion of parasites, and mitochondrial fusion (Lemberg and Freeman 2007). High-resolution structures of bacterial rhomboid proteases have recently provided insight into the mechanism of intramembrane proteolysis by serine ICLIPs (Ben-Shem et al. 2007; Lemieux et al. 2007; Wang et al. 2006b; Wu et al. 2006). An intramembranous active site Ser-His dyad as well as the presence of water within a hydrophilic cavity formed by the TMDs have been demonstrated (Lemberg and Freeman 2007). This finding therefore provided the ultimate and unequivocal proof that proteolysis within the hydrophobic bilayer of the membrane is possible. Interestingly, the family of rhomboid proteases seems to be the only ICLIP class that does not necessarily require an initial shedding event preceding the intramembrane cleavage.

The class of GxGD-type aspartyl proteases (Fig. 1) so far covers three ICLIP families, the presenilins (PS), known to be involved in the pathogenesis of AD, the family of signal peptide peptidase (SPP) and SPP-like proteases (SPPL), and the family of bacterial type IV prepelin peptidases (TFPPs; Friedmann et al. 2004; LaPointe and Taylor 2000; Ponting et al. 2002; Steiner et al. 2000; Weihofen et al. 2002).

The PS and SPP/SPPL families share a lot of similarities, but fundamental differences regarding their localization, their molecular composition and their cellular function have also been recently discovered.

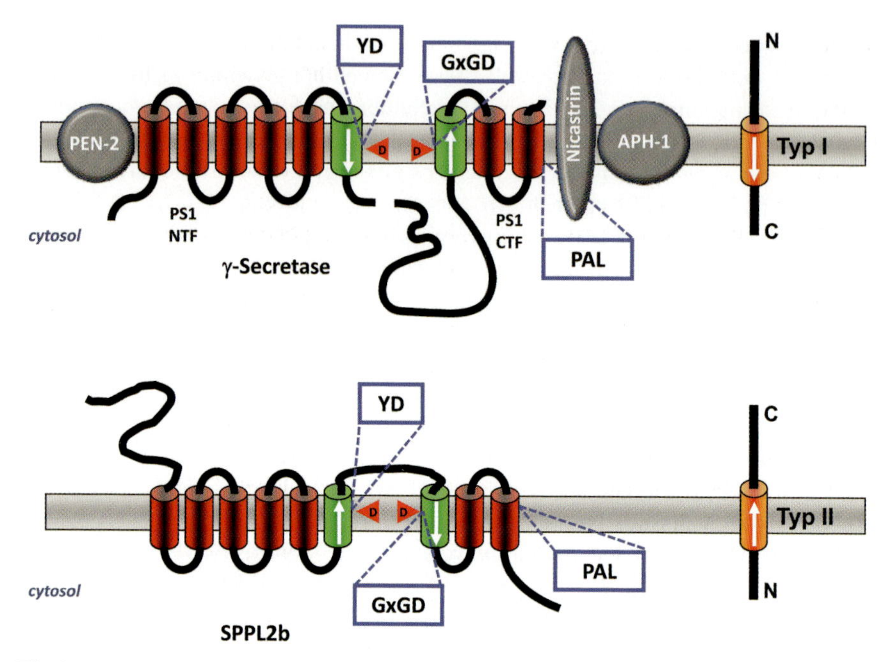

Fig. 2 Model depicting the γ-secreatse complex and SPPL2b and the orientation of their substrates. The conserved motifs contributing to the active site of the proteases are highlighted. Note that the active site domains of the two enzymes are oriented in exactly the opposite way

In this chapter we will compare the biochemical, functional and structural properties of these protease families with a strong focus on AD γ-secretase and SPPL2b, the best-characterized member of the SPPL subfamily (Fig. 2).

2 Structural and Molecular Organization of Intramembrane Cleaving GxGD-type Aspartyl Proteases

Although PSs were the founding members of the class of GxGD-type aspartyl proteases, they turned out to be the most complicated family. While the PSs, which provide the active center of γ-secretase, are members of a high molecular weight complex (Haass and Steiner 2002), SPP/SPPLs and TFPPs seem to act as dimers or even only monomers (LaPointe and Taylor 2000; Weihofen et al. 2002). In addition to PS, the γ-secretase complex contains three other essential integral membrane proteins: nicastrin (NCT), anterior pharynx defective 1 (APH-1) and presenilin enhancer 2 (PEN-2; Francis et al. 2002; Goutte et al. 2002; Fig. 2). NCT, a $\sim$ 100 kDa type I transmembrane glycoprotein carrying a large ectodomain and a short cytoplasmic domain (Yu et al. 2000), probably serves as γ-secretase substrate receptor (Shah et al. 2005). PEN-2 is required for the stabilization of the autocatalytically generated PS fragments (Thinakaran et al. 1996) in the complex (Hasegawa et al. 2004; Prokop et al. 2004), whereas the function of APH-1 is still unclear. Together with the $\sim$ 50 kDa PS, the components form a complex of roughly 500 kDa, implying that each component may be represented twice within the complex. Whether γ-secretase indeed needs to form a dimer to be active or whether a single complex by itself provides proteolytic activity is currently under debate (Sato et al. 2007). The absolute requirement of these four components to form an active γ-secretase complex was proven by the reconstitution of γ-secretase in yeast (Edbauer et al. 2003). Only upon expression of all four γ-secretase complex components proteolytic activity is achieved; overexpression of PS alone is not sufficient. In contrast, to obtain increased SPP/SPPL activity, it is sufficient to simply overexpress the protease (Fluhrer et al. 2006; Friedmann et al. 2006; Nyborg et al. 2004a), indicating that SPP/SPPLs do not need any other essential co-factors for proteolytic activity (Fig. 2). There is evidence that SPP as well as SPPLs form homodimers (Friedmann et al. 2004a, b). The homodimer was selectively labeled by an active site inhibitor, strongly supporting the notion that dimerization is required for biological activity. However, in a later study using a different inhibitor, selective labeling of the monomer was observed (Sato et al. 2006). Whether SPP/SPPLs under physiological conditions have additional transient interactors positively or negatively regulating their proteolytic activity needs to be investigated. For γ-secretase, CD147 and TMP21 are proposed to fulfill such a regulatory activity (Chen et al. 2006; Zhou et al. 2005), although a very recent observation suggests that CD147 does not directly interact with γ-secretase but rather modulates extracellular degradation of Aβ (Vetrivel et al. 2008).

While SPP/SPPLs are active as full-length proteins, PS undergoes endoproteolysis (Thinakaran et al. 1996). This endoproteolytic cleavage is most likely an autoproteolytic event (Edbauer et al. 2003; Wolfe et al. 1999); however, this has not been directly proven.

The catalytic center of GxGD-type aspartyl proteases contains two critical aspartate residues located within the two neighboring TMDs 6 and 7 of the protein (Fig. 2). The N-terminal catalytically active site aspartate is embedded in a conserved YD motif, whereas the C-terminal active site domain contains the equally conserved GxGD motif (Steiner et al. 2000; Wolfe et al. 1999). While the catalytic motifs of PSs and SPP/SPPLs are likely located within the hydrophobic core of TM6 and TM7, the active site of the bacterial TFFPs is most likely located at the cytoplasmic border and probably not within the membrane (LaPointe and Taylor 2000). Mutagenesis of either critical aspartate residue in PSs, SPP and TFFPs abolishes their proteolytic activity (LaPointe and Taylor 2000; Weihofen et al. 2002; Wolfe et al. 1999). In zebrafish, expression of GxGD aspartate mutants of SPP/SPPLs phenocopy a morpholino-mediated, knockdown phenotype of the respective SPP/SPPL family member (Krawitz et al. 2005). The formal proof of the requirement of the aspartate within the YD motif of SPPL family members is still missing. The glycine directly N-terminal to the aspartate within the GxGD motif is also required for proteolytic activity of GxGD-aspartyl proteases. In PS1 and SPPL2b, the only other amino acid tolerated at this position is an alanine. Nonetheless the substrate conversion of SPPL2b carrying the G/A mutation is significantly slower compared to the wt enzyme (Fluhrer et al. 2008). PS carrying the G/A mutation strongly affects the Aβ 42/40 ratio by selectively lowering Aβ 40 production (Steiner et al. 2000; Fluhrer et al. 2008). The function of the Y within the YD motif of SPP/SPPLs and TFPPs has not been investigated so far, but it is known that, for example, the mutation YD/SD in PS1 causes early onset familial AD (FAD) (Miklossy et al. 2003 and *www.molgen.ua.ac.be/ADMutations*). So it is tempting to speculate, that like the glycine in close vicinity to the aspartate in the GxGD motif, the tyrosine in the YD motif is required for proper function of the enzyme. At least in PSs and SPP/SPPL family members, a third highly conserved motif, likely to contribute to the active center of GxGD-type aspartyl proteases, is found. The so-called PAL sequence is located in the most C-terminal TMD of GxGD-type aspartyl proteases. The important participation of the PAL motif in the catalytic center is supported by the finding that a transition-state analog inhibitor fails to bind to SPP and PS upon mutagenesis of the PAL sequence (Wang et al. 2006a). It is currently unknown how the PAL domain affects PS and SPP activity; however, one may assume a close proximity of the TMDs.

3 Cellular Localization

Originally it was believed that PSs were exclusively localized to early secretory compartments like the endoplasmic reticulum (ER) and the intermediate compartment (Annaert and De Strooper 1999; Cupers et al. 2001). These findings created a large debate in the field, since γ-secretase activity per se was believed to take place at the cell surface (Haass et al. 1993). This phenomenon is known in the literature as the "spatial paradox" (Checler 2001; Cupers et al. 2001). With the identification of Nicastrin as a component of the γ-secretase complex (Yu et al. 2000), it was shown that the γ-secretase complex assembles in the ER and is then targeted through the secretory pathway, where Nicastrin becomes endoglycosidase H-resistant (Kaether et al. 2002). Cell-surface biotinylation assays and live cell microscopy further demonstrate that a small but fully active amount of γ-secretase is localized to the cell surface (Kaether et al. 2002), whereas a majority of unincorporated PS1 is retained in the ER (Capell et al. 2005; Kaether et al. 2004). Recently, a first protein factor, namely Rer1 (Retention in the endoplasmic reticulum 1), was shown to be required for γ-secretase complex formation or retention of PEN-2 within the ER (Annaert et al. 1999; Kaether et al. 2007).

SPP is exclusively detected in the ER; (Friedmann et al. 2006; Krawitz et al. 2005; Fig. 3), accompanied by the substrate preference of SPP that cleaves signal peptides of proteins translated into the ER (Weihofen et al. 2002). Interestingly, although sharing a high sequence homology, SPP and SPPLs localize to different cellular compartments (Fig. 3). SPPL2a and b accumulate on the plasma membrane and within endosomal/lysosomal compartments (Friedmann et al. 2006; Krawitz et al. 2005; Fig. 3). SPPL3 has been detected in the ER (Krawitz et al. 2005;

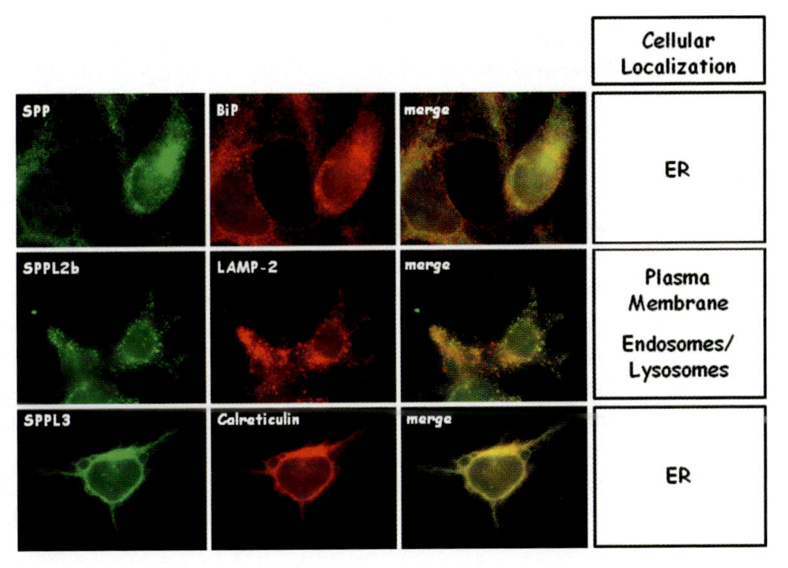

Fig. 3 Differential localization of SPP/SPPL family members. Immunofluorescence staining of SPP, SPPL2b and SPPL3 reveals endoplasmic reticulum (ER) localization for SPP and SPPL3; SPPL2b predominantly localizes to later secretory compartments, including endosomes/lysosomes

Fig. 3) as well as within later compartments (Friedmann et al. 2006). Since SPP only cleaves substrates located in the ER membrane (Lemberg and Martoglio 2002; Weihofen et al. 2002) and all known substrates for SPPL2a and SPPL2b are targeted to the cell surface (Fluhrer et al. 2006; Friedmann et al. 2006; Kirkin et al. 2007; Martin et al. 2008), the substrate selection of SPP/SPPLs may be achieved by their differential subcellular localization. How the distinct cellular localization of SPP/SPPLs is achieved is not yet entirely clear, but SPP may be actively retained within the ER by its putative KKXX retention signal (Weihofen et al. 2002), which is not present in any of the members of the SPPL family.

4 Substrate Requirements and Physiological Function

Members of the SPP/SPPL family apparently only accept single pass transmembrane proteins of type II orientation as substrates, whereas PSs exclusively recognize type I trans-membrane proteins (Fig. 2). Since both protease families seem to have numerous substrates, it is discussed that GxGD-type aspartyl proteases fulfill the function of a so-called membrane proteasome (Kopan and Ilagan 2004), removing the sticky transmembrane domains from the cellular membranes that are left behind after proteolytic processing of transmembrane proteins, e.g., shedding of ligands or receptors at the cell surface. How PSs and SPP/SPPLs are able to discriminate between type I and type II substrates is currently not fully understood. Strikingly, the active site domains in PSs and SPP/SPPLs are predicted to be arranged in exactly opposite orientations (Weihofen and Martoglio 2003; Fig. 2), which might reflect the opposite orientation of the substrates. Another possibility for substrate discrimination is the receptor proteins or domains within the γ-secretase complex and SPP/SPPL. The initial recognition of γ-secretase substrates requires the main part of the substrate ectodomain to be removed by shedding. The γ-secretase substrates are then recognized by NCT, which identifies the free N-terminus of the substrate (Shah et al. 2005). Therefore. shedding is a prerequisite for every physiological γ-secretase substrate. Since SPP/SPPLs do not require any co-factors for activity (Fig. 2), the receptor for substrate recognition must be located within SPP/SPPLs themselves, but a defined domain for the substrate recognition has not yet been identified. Maybe the active site itself is involved in substrate recognition, as has recently been shown for PS1, where the active site domain overlaps with a second substrate recognition site (Kornilova et al. 2005; Yamasaki et al. 2006). Shedding of type-II proteins seems to greatly facilitate intramembrane proteolysis of SPPL substrates. In contrast to γ-secretase substrates, shedding is not an absolute prerequisite for intramembrane proteolysis (Martin, Fluhrer and Haass, unpublished data). This may reflect the absence of NCT as a docking protein involved in substrate identification.

SPP predominantly cleaves signal peptides that are removed from the nascent protein chain by signal-peptidase (SP) in the ER (Fig. 1), in the middle of their hydrophobic core. All signal peptides adopt a type II orientation during co-translation and are therefore, in principle, preferred substrates of SPP (Weihofen et al. 2002). But although a variety of signal peptides from human and viral proteins - like the hormone prolactin, MHC class I molecules and calreticulin - are cleaved by SPP, examples of signal peptides that are not substrates for SPP, like that of RNAse A and human cytomegalovirus glycoprotein UL40, have been published (Lemberg and Martoglio 2002). Therefore, another protease with SPP function might exist, at least in humans. Potential candidate proteins would be the SPP homologues, SPPL3 and SPPL2c, which may both localize to the ER (Friedmann et al. 2006; Krawitz et al. 2005). But so far no substrates have been described for these proteases. Interestingly, however, the knockdown phenotypes of the SPP and the SPPL3 homologue in zebrafish result in virtually indistinguishable phenotypes (Krawitz et al. 2005), which might point to a similar cellular function for the two proteases.

The best-understood γ-secretase substrates are βAPP and the Notch receptor, Notch 1. Although the intramembrane proteolysis of βAPP contributes to the production of Aβ (Fig. 4) and therefore to the pathogenesis of AD, the intramembrane proteolysis of Notch 1 is of much greater physiological relevance. This is reflected by the fact that ablation of PS1/2 and other γ-secretase complex components in many different organisms results in a lethal Notch phenotype (Selkoe and Kopan 2003). The Notch receptors are known to bind ligands like Serrate and Jagged on the cell surface (Selkoe and Kopan 2003). Upon ligand binding, the receptor/ligand-complex starts being endocytosed by the ligand-expressing cell, inducing shedding of the receptor by ADAM proteases (Gordon et al. 2007). On the

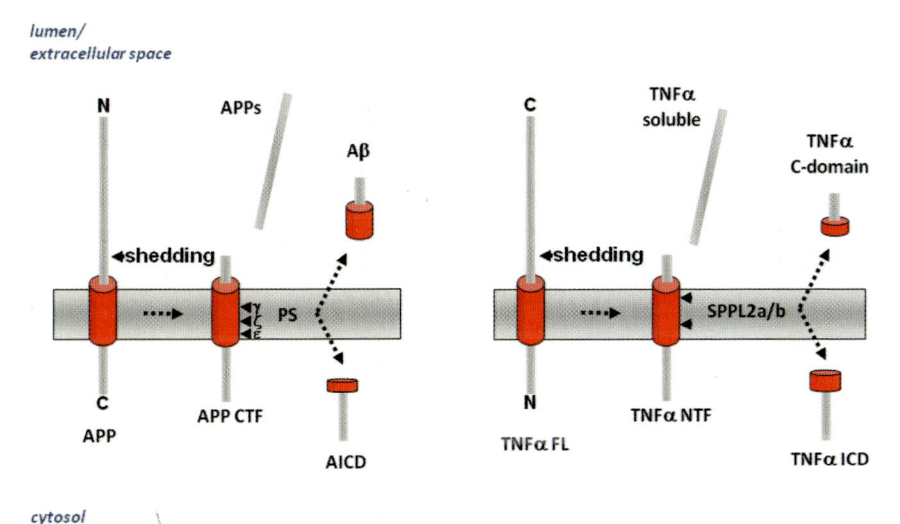

Fig. 4 Model showing RIP of APP and TNFα. The individual cleavage products after shedding and intramembrane proteolysis are depicted

receptor-expressing cell, γ-secretase cleaves the remaining stub of the Notch receptor, liberating the Notch ICD, which has been shown to translocate to the nucleus regulating gene transcription (Selkoe and Kopan 2003).

Recently, tumor necrosis factor α (TNFα; Fluhrer et al. 2006; Friedmann et al. 2006), the FAS ligand (FasL; Kirkin et al. 2007) and Bri2 (Itm2b; Martin et al. 2008) have been identified as substrates for intramembrane proteolysis by SPPL2a and SPPL2b. All three substrates, like Notch and APP, undergo shedding by a protease of the ADAM family (Fluhrer et al. 2006; Friedmann et al. 2006; Kirkin et al. 2007; Martin et al. 2008). TNFα is a well-known, pro-inflammatory cytokine that has a critical role in autoimmune disorders such as rheumatoid arthritis and Crohn's disease (Locksley et al. 2001; Vassalli 1992). These effects are mediated by the ectodomain of TNFα (TNFα soluble), which is released by TACE/ADAM17 from the cell surface of the TNFα-expressing cell (Hooper et al. 1997; Schlondorff and Blobel 1999; Fig. 4). The TNFα ectodomain then enters the blood stream (Gearing et al. 1994; McGeehan et al. 1994) and binds to a variety of different receptors on the signal-receiving cell, triggering the respective signal cascade. In the signal-sending cell, the TNFα stub (TNFα NTF) is left behind (Fig. 4). This TNFα NTF is substrate to intramembrane proteolysis by SPPL2a and SPPL2b (Fluhrer et al. 2006; Friedmann et al. 2006), releasing a short TNFα ICD to the cytosol and the corresponding part, the TNFα C-domain, to the extracellular/luminal space of the cell (Fluhrer et al. 2006; Fig. 4). The ICD of TNFα has been shown to stimulate expression of interleukin-12 in the signal-sending cell (Friedmann et al. 2006), a mechanism that is referred to as TNFα reverse signaling. Similarly, the ICD of the FasL translocates to the nucleus, where it may act as a suppressor of gene transcription (Kirkin et al. 2007). No physiological function has yet been assigned for the corresponding Bri2 ICD.

Interestingly, a variety of the RIP substrates, like APP, Bri2 and TNFα, have been shown to dimerize or even trimerize (Kriegler et al. 1988; Munter et al. 2007; Tsachaki et al. 2008). The dimerization of APP and Bri2 is mediated by GxxxG dimerization motifs (Munter et al. 2007; Tsachaki et al. 2008) that, when disrupted, lead to altered intramembrane cleavage, at least in the case of APP (Munter et al. 2007). However, the precise mechanism of how the GxxxG motif triggers intramembrane proteolysis is currently unclear.

5 Cleavage Mechanism

Mostly, all TMDs of RIP substrates are predicted to adopt an α-helical confirmation. Therefore, it is likely that the TMDs require unwinding prior to the occurrence of endoproteolysis. For SPP substrates, this is proposed to be promoted by helix-breaking residues within the hydrophobic core of the signal peptides (Lemberg and Martoglio 2002). Mutation of such residues in the TMD of TNFα, on the other hand, does not significantly affect intramembrane proteolysis by SPPL2b (Fluhrer and Haass, unpublished data). Unfortunately, structural information is not available for any of the human SPP family members. Therefore, it is difficult to predict how exactly the intramembrane cleavage is mechanistically performed. However, the N-terminus of the secreted TNFα C-domain and the C-terminus of the TNFα ICD do not exactly match, but are rather separated by 10–15 amino acids (Fluhrer et al. 2006; Fig. 4). Maybe the unwinding of the α-helical substrate conformation is facilitated by multiple cleavage events, which step by step open the α-helix like an elastic spring. Consequently, these multiple cleavage events allow efficient release of hydrophobic TMDs from cellular membranes. If such multiple cleavages occur with other SPP/SPPL substrates as well is currently unknown. However, a synthetic substrate for SPP has been shown to undergo one major as well as several other minor intramembrane cuts (Sato et al. 2006). Like for TNFα, multiple cleavage events have been reported specifically for the γ-secretase substrates APP and Notch (Fluhrer et al. 2008), further supporting the idea of unwinding the α-helix of the substrate with every individual cleavage. Once the substrate has accessed the catalytic site of γ-secretase, it is cleaved within its TMD at three topologically distinct sites, termed ε-, ζ- and γ-sites (Haass and Selkoe 2007). In the case of APP, the first cleavage at the ε-site releases the APP intracellular domain (AICD; Fig. 4) into the cytosol. The remaining part of the APP TMD is further cleaved at the ζ- and γ-sites until Aβ is short enough to be released from the membrane (Fig. 4). Interestingly, the cleavages at the ε-, ζ- and γ-sites are heterogeneous, suggesting the existence of two different product lines, leading to the benign Aβ40 on the one hand and to the pathogenic Aβ42 on the other hand (Qi-Takahara et al. 2005). The pathogenic product line generating Aβ42 seems to be dominant in some but not all PS FAD mutants (Qi-Takahara et al. 2005). FAD mutations seem to directly or indirectly affect the confirmation of the active site of γ-secretase, selectively slowing the product line leading to the benign Aβ40 and therefore causing a relative increase of the pathologic Aβ42, leading to early onset AD (Fluhrer et al. 2008). When a FAD mutation was transferred to the corresponding site in SPPL2b, the sequential cleavage of TNFα was similarly slowed (Fluhrer et al. 2008). However, the precise mechanism of sequential processing by intramembrane GxGD aspartyl proteases is still unclear.

6 Inhibition of Intramembrane Cleaving Aspartyl Proteases: A Therapeutic Target

Since substrates of intramembrane proteases are frequently involved in the development of diseases (see above), the proteases processing these substrates are drug targets. Inhibition of γ-secretase activity, for example, is an important approach for therapeutic treatment of AD and γ-secretase inhibitor identification and development reaches an advanced state (Churcher and Beher 2005). Unfortunately, γ-secretase inhibitors not only block the processing of APP, avoiding the production of Aβ, but also interfere with Notch signaling. Therefore, γ-secretase inhibitors affect cellular differentiation and cause severe side effects. Moreover, active site γ-secretase inhibitors have been shown to cross react with SPP (Iben et al. 2007; Weihofen et al. 2003) and are likely to also block the members of the SPPL family, since the active site of the GxGD proteases is highly conserved.

Therefore, the development of selective inhibitors is a major challenge for the pharmaceutical industry and academic institutions. Besides synthetic γ-secretase inhibitors, a very well-known class of non-steroidal anti-inflammatory drugs (NSAIDs) has been shown to selectively decrease cleavage of γ-secretase at the γ-42 site of APP without affecting cleavage at the γ-40 and the ε-site (Weggen et al. 2001). Thus NSAIDS do not affect the γ-secretase-mediated release of NICD from Notch (Weggen et al. 2001). Whether these NSAIDs directly bind γ-secretase or APP is currently under debate (Beher et al. 2004; Takahashi et al. 2003). Interestingly, NSAIDs also seem to affect the proteolytic activity of SPP (Sato et al. 2006), either pointing to a direct binding to the enzyme or to a more general mechanism, such as an effect on the lipid composition of the membrane and, therefore, indirectly affecting the conformation of GxGD proteases.

7 Conclusions

We have described and compared the biochemical and cellular properties of GxGD-type aspartyl proteases. Although we observed some fundamental differences in terms of substrate recognition and orientation, as well as the requirement of shedding and the role of essential co-factors, the molecular mechanisms of intramembrane proteolysis appear to be surprisingly similar. Multiple intramembrane cleavages are performed to release small peptides and to finally remove the TMD from the cellular membranes. FAD-associated mutations affect the kinetics of these intramembrane cleavages. In the case of APP processing specifically, the production of the benign Aß40 is reduced whereas the generation of the neurotoxic Aβ42 remains unaffected. A similar partial loss of function was observed when a FAD-like mutation occurring in PS1 was introduced at a homologous position of SPPL2b. Our findings, therefore, finally provide a solution for the long-lasting debate over whether PS mutations cause a loss or a gain of function. Based on the evidence presented, these mutations cause a partial loss of function. This finding may have important therapeutic implications, since treatment of patients with low concentrations of γ-secretase inhibitors (used to avoid an inhibition of Notch signaling) may lead to a selective reduction of Aß40 and thus increase the Aß42/40 ratio. Moreover, care must be taken to avoid cross-reactivity of γ-secretase inhibitors with the homologous SPP/SPPL.

Acknowledgements This work is supported by the Deutsche Forschungsgemeinschaft (Gottfried Wilhelm Leibniz-Award (to C.H.) and HA 1737-11 (to C.H. and R.F.)) and the NGFN-2. The LMU excellent program supports C.H. with a research professorship.

References

Annaert W, De Strooper B (1999) Presenilins: molecular switches between proteolysis and signal transduction. Trends Neurosci 22: 439–443

Annaert WG, Levesque L, Craessaerts K, Dierinck I, Snellings G, Westaway D, George-Hyslop PS, Cordell B, Fraser P, De Strooper B (1999) Presenilin 1 controls γ-secretase processing of amyloid precursor protein in pre-golgi compartments of hippocampal neurons. J Cell Biol 147: 277–294

Beher D, Clarke EE, Wrigley JD, Martin AC, Nadin A, Churcher I, Shearman MS (2004) Selected non-steroidal anti-inflammatory drugs and their derivatives target gamma-secretase at a novel site. Evidence for an allosteric mechanism. J Biol Chem 279: 43419–43426

Ben-Shem A, Fass D, Bibi E (2007) Structural basis for intramembrane proteolysis by rhomboid serine proteases. Proc Natl Acad Sci USA 104: 462–466

Brown MS, Goldstein JL (1999) A proteolytic pathway that controls the cholesterol content of membranes, cells, and blood. Proc Natl Acad Sci USA 96: 11041–11048.

Capell A, Beher D, Prokop S, Steiner H, Kaether C, Shearman MS, Haass C (2005) Gamma-secretase complex assembly within the early secretory pathway. J Biol Chem 280: 6471–6478

Checler F (2001) The multiple paradoxes of presenilins. J Neurochem 76: 1621–1627

Chen F, Hasegawa H, Schmitt-Ulms G, Kawarai T, Bohm C, Katayama T, Gu Y, Sanjo N, Glista M, Rogaeva E, Wakutani Y, Pardossi-Piquard R, Ruan X, Tandon A, Checler F, Marambaud P, Hansen K, Westaway D, St George-Hyslop P, Fraser P (2006) TMP21 is a presenilin complex component that modulates gamma-secretase but not epsilon-secretase activity. Nature 440: 1208–1212

Churcher I, Beher D (2005) Gamma-secretase as a therapeutic target for the treatment of Alzheimer's disease. Curr Phamaceut Design 11: 3363–3382

Cupers P, Bentahir M, Craessaerts K, Orlans I, Vanderstichele H, Saftig P, De Strooper B, Annaert W (2001) The discrepancy between presenilin subcellular localization and γ-secretase processing of amyloid precursor protein. J Cell Biol 154: 731–740

Edbauer D, Winkler E, Regula JT, Pesold B, Steiner H, Haass C (2003) Reconstitution of γ-secretase activity. Nature Cell Biol 5: 486–488

Feng L, Yan H, Wu Z, Yan N, Wang Z, Jeffrey PD, Shi Y (2007) Structure of a site-2 protease family intramembrane metalloprotease. Science 318: 1608–1612

Fluhrer R, Grammer G, Israel L, Condron MM, Haffner C, Friedmann E, Bohland C, Imhof A, Martoglio B, Teplow DB, Haass C (2006) A gamma-secretase-like intramembrane cleavage of TNFalpha by the GxGD aspartyl protease SPPL2b. Nature Cell Biol 8: 894–896

Fluhrer R, Steiner H, Haass C (2008) Intramembrane Proteolysis by gamma-Secretase and related GxGD-type Aspartylproteases. J Biol Chem, in press

Fluhrer R, Fukumori A, Martin L, Grammer G, Haug-Kröper M, Klier B, Winkler E, Kremmer E, Condron MM, Teplow DB, Steiner H, Haass C (2008) Intramembrane proteolysis of G×GD-type aspartyl proteases is slowed by a familial Alzheimer disease-like mutation. J Biol Chem, in press

Francis R, McGrath G, Zhang J, Ruddy DA, Sym M, Apfeld J, Nicoll M, Maxwell M, Hai B, Ellis MC, Parks AL, Xu W, Li J, Gurney M, Myers RL, Himes CS, Hiebsch RD, Ruble C, Nye JS, Curtis D (2002) aph-1 and pen-2 are required for Notch pathway signaling, γ-secretase cleavage of βAPP, and presenilin protein accumulation. Dev Cell 3: 85–97

Friedmann E, Lemberg MK, Weihofen A, Dev KK, Dengler U, Rovelli G, Martoglio B (2004) Consensus analysis of signal peptide peptidase and homologous human aspartic proteases reveals opposite topology of catalytic domains compared with presenilins. JBbiol Chem 279: 50790–50798

Friedmann E, Hauben E, Maylandt K, Schleeger S, Vreugde S, Lichtenthaler SF, Kuhn PH, Stauffer D, Rovelli G, Martoglio B (2006) SPPL2a and SPPL2b promote intramembrane proteolysis of TNFalpha in activated dendritic cells to trigger IL-12 production. Nature Cell Biol 8: 843–848

Gearing AJ, Beckett P, Christodoulou M, Churchill M, Clements J, Davidson AH, Drummond AH, Galloway WA, Gilbert R, Gordon JL, Leber TM, Mangan M, Miller K, Nayee P, Owen K, Patel S, Thomas W, Wells G, Wood LM, Wooley K (1994) Processing of tumour necrosis factor-alpha precursor by metalloproteinases. Nature 370: 555–557

Gordon WR, Vardar-Ulu D, Histen G, Sanchez-Irizarry C, Aster JC, Blacklow SC (2007) Structural basis for autoinhibition of Notch. Nature Struct Mol Biol 14: 295–300

Goutte C, Tsunozaki M, Hale VA, Priess JR (2002) APH-1 is a multipass membrane protein essential for the Notch signaling pathway in Caenorhabditis elegans embryos. Proc Natl Acad Sci USA 99: 775–779.

Haass C (2004) Take five-BACE and the γ-secretase quartet conduct Alzheimer's amyloid β-peptide generation. EMBO J 23: 483–488

Haass C, Selkoe DJ (2007) Soluble protein oligomers in neurodegeneration: lessons from the Alzheimer's amyloid beta-peptide. Nature Rev Mol Cell Biol 8: 101–112

Haass C, Steiner H (2002) Alzheimer disease γ-secretase: a complex story of GxGD-type presenilin proteases. Trends Cell Biol 12: 556–562

Haass C, Hung AY, Schlossmacher MG, Teplow DB, Selkoe DJ (1993) beta-Amyloid peptide and a 3-kDa fragment are derived by distinct cellular mechanisms. J Biol Chem 268: 3021–3024

Hardy J, Selkoe DJ (2002) The amyloid hypothesis of Alzheimer's disease: progress and problems on the road to therapeutics. Science 297: 353–356

Hasegawa H, Sanjo N, Chen F, Gu YJ, Shier C, Petit A, Kawarai T, Katayama T, Schmidt SD, Mathews PM, Schmitt-Ulms G, Fraser PE, St George-Hyslop P (2004) Both the sequence and length of the C terminus of PEN-2 are critical for intermolecular interactions and function of presenilin complexes. J Biol Chem 279: 46455–46463

Hooper NM, Karran EH, Turner AJ (1997) Membrane protein secretases. Biochem J 321 (Pt 2): 265–279

Iben LG, Olson RE, Balanda LA, Jayachandra S, Robertson BJ, Hay V, Corradi J, Prasad CV, Zaczek R, Albright CF, Toyn JH (2007) Signal peptide peptidase and gamma-secretase share equivalent inhibitor binding pharmacology. J Biol Chem 282: 36829–36836

Kaether C, Lammich S, Edbauer D, Ertl M, Rietdorf J, Capell A, Steiner H, Haass C (2002) Presenilin-1 affects trafficking and processing of betaAPP and is targeted in a complex with nicastrin to the plasma membrane. J Cell Biol 158: 551–561

Kaether C, Capell A, Edbauer D, Winkler E, Novak B, Steiner H, Haass C (2004) The presenilin C-terminus is required for ER-retention, nicastrin-binding and gamma-secretase activity. EMBO J 23: 4738–4748

Kaether C, Scheuermann J, Fassler M, Zilow S, Shirotani K, Valkova C, Novak B, Kacmar S, Steiner H, Haass C (2007) Endoplasmic reticulum retention of the gamma-secretase complex component Pen2 by Rer1. EMBO Rep 8: 743–748

Kirkin V, Cahuzac N, Guardiola-Serrano F, Huault S, Luckerath K, Friedmann E, Novac N, Wels WS, Martoglio B, Hueber AO, Zornig M (2007) The Fas ligand intracellular domain is released by ADAM10 and SPPL2a cleavage in T-cells. Cell Death Differentiation 14: 1678–1687

Kopan R, Ilagan MX (2004) Gamma-secretase: proteasome of the membrane? Nature Rev Mol Cell Biol 5: 499–504

Kornilova AY, Bihel F, Das C, Wolfe MS (2005) The initial substrate-binding site of gamma-secretase is located on presenilin near the active site. Proc Natl Acad Sci USA102: 3230–3235

Krawitz P, Haffner C, Fluhrer R, Steiner H, Schmid B, Haass C (2005) Differential localization and identification of a critical aspartate suggest non-redundant proteolytic functions of the presenilin homologues SPPL2b and SPPL3. J Biol Chem 280: 39515–39523

Kriegler M, Perez C, DeFay K, Albert I, Lu SD (1988) A novel form of TNF/cachectin is a cell surface cytotoxic transmembrane protein: ramifications for the complex physiology of TNF. Cell 53: 45–53

LaPointe CF, Taylor RK (2000) The type 4 prepilin peptidases comprise a novel family of aspartic acid proteases. J Biol Chem 275: 1502–1510

Lazarov VK, Fraering PC, Ye W, Wolfe MS, Selkoe DJ, Li H (2006) From the Cover: Electron microscopic structure of purified, active {gamma}-secretase reveals an aqueous intramembrane chamber and two pores. Proc Natl Acad Sci USA 103: 6889–6894

Lemberg MK, Martoglio B (2002) Requirements for signal peptide peptidase-catalyzed intramembrane proteolysis. Mol Cell 10: 735–744

Lemberg MK, Freeman M (2007) Functional and evolutionary implications of enhanced genomic analysis of rhomboid intramembrane proteases. Genome Res 17: 1634–1646

Lemieux MJ, Fischer SJ, Cherney MM, Bateman KS, James MN (2007) The crystal structure of the rhomboid peptidase from Haemophilus influenzae provides insight into intramembrane proteolysis. Proc Natl Acad Sci USA104: 750–754

Locksley RM, Killeen N, Lenardo MJ (2001) The TNF and TNF receptor superfamilies: integrating mammalian biology. Cell 104: 487–501

Martin L, Fluhrer R, Reiss K, Kremmer E, Saftig P, Haass C (2008) Regulated intramembrane proteolysis of Bri2 (Itm2b) by ADAM10 and SPPL2a/SPPL2b. J Biol Chem 283: 1644–1652

McGeehan GM, Becherer JD, Bast RC, Jr., Boyer CM, Champion B, Connolly KM, Conway JG, Furdon P, Karp S, Kidao S, et al. (1994) Regulation of tumour necrosis factor-alpha processing by a metalloproteinase inhibitor. Nature 370: 558–561

Miklossy J, Taddei K, Suva D, Verdile G, Fonte J, Fisher C, Gnjec A, Ghika J, Suard F, Mehta PD, McLean CA, Masters CL, Brooks WS, Martins RN (2003) Two novel presenilin-1 mutations (Y256S and Q222H) are associated with early-onset Alzheimer's disease. Neurobiol Aging 24: 655–662

Munter LM, Voigt P, Harmeier A, Kaden D, Gottschalk KE, Weise C, Pipkorn R, Schaefer M, Langosch D, Multhaup G (2007) GxxxG motifs within the amyloid precursor protein transmembrane sequence are critical for the etiology of Abeta42. EMBO J 26: 1702–1712

Nyborg AC, Jansen K, Ladd TB, Fauq A, Golde TE (2004a) A signal peptide peptidase (SPP) reporter activity assay based on the cleavage of type II membrane protein substrates provides further evidence for an inverted orientation of the SPP active site relative to presenilin. J Biol Chem 279: 43148–43156

Nyborg AC, Kornilova AY, Jansen K, Ladd TB, Wolfe MS, Golde TE (2004b) Signal peptide peptidase forms a homodimer that is labeled by an active site-directed gamma-secretase inhibitor. J Biol Chem 279: 15153–15160

Ponting CP, Hutton M, Nyborg A, Baker M, Jansen K, Golde TE (2002) Identification of a novel family of presenilin homologues. Human Mol Genet 11: 1037–1044

Prokop S, Shirotani K, Edbauer D, Haass C, Steiner H (2004) Requirement of PEN-2 for stabilization of the presenilin N-/C-terminal fragment heterodimer within the gamma-secretase complex. J Biol Chem 279: 23255–23261

Qi-Takahara Y, Morishima-Kawashima M, Tanimura Y, Dolios G, Hirotani N, Horikoshi Y, Kametani F, Maeda M, Saido TC, Wang R, Ihara Y (2005) Longer forms of amyloid beta protein: implications for the mechanism of intramembrane cleavage by gamma-secretase. J Neurosci 25: 436–445

Rawson RB, Zelenski NG, Nijhawan D, Ye J, Sakai J, Hasan MT, Chang TY, Brown MS, Goldstein JL (1997) Complementation cloning of S2P, a gene encoding a putative metalloprotease required for intramembrane cleavage of SREBPs. Mol Cell 1: 47–57.

Sato T, Nyborg AC, Iwata N, Diehl TS, Saido TC, Golde TE, Wolfe MS (2006) Signal peptide peptidase: biochemical properties and modulation by nonsteroidal antiinflammatory drugs. Biochemistry 45: 8649–8656

Sato T, Diehl TS, Narayanan S, Funamoto S, Ihara Y, De Strooper B, Steiner H, Haass C, Wolfe MS (2007) Active gamma-secretase complexes contain only one of each component. J Biol Chem 282: 33985–33993

Schlondorff J, Blobel CP (1999) Metalloprotease-disintegrins: modular proteins capable of promoting cell-cell interactions and triggering signals by protein-ectodomain shedding. J Sell Sci 112 (Pt 21): 3603–3617

Selkoe D, Kopan R (2003) Notch and Presenilin: regulated intramembrane proteolysis links development and degeneration. Annu Rev Neurosci 26: 565–597

Shah S, Lee SF, Tabuchi K, Hao YH, Yu C, LaPlant Q, Ball H, Dann CE, 3rd, Sudhof T, Yu G (2005) Nicastrin functions as a gamma-secretase-substrate receptor. Cell 122: 435–447

Steiner H, Kostka M, Romig H, Basset G, Pesold B, Hardy J, Capell A, Meyn L, Grim MG, Baumeister R, Fechteler K, Haass C (2000) Glycine 384 is required for presenilin-1 function and is conserved in polytopic bacterial aspartyl proteases. Nature Cell Biol 2: 848–851

Steiner H, Than M, Bode W, Haass C (2006) Pore-forming scissors? A first structural glimpse of gamma-secretase. Trends Biochem Sci 31: 491–493

Takahashi Y, Hayashi I, Tominari Y, Rikimaru K, Morohashi Y, Kan T, Natsugari H, Fukuyama T, Tomita T, Iwatsubo T (2003) Sulindac sulfide is a noncompetitive gamma-secretase inhibitor that preferentially reduces Abeta 42 generation. J Biol Chem 278: 18664–18670

Thinakaran G, Borchelt DR, Lee MK, Slunt HH, Spitzer L, Kim G, Ratovitsky T, Davenport F, Nordstedt C, Seeger M, Hardy J, Levey AI, Gandy SE, Jenkins NA, Copeland NG, Price DL, Sisodia SS (1996) Endoproteolysis of presenilin 1 and accumulation of processed derivatives in vivo. Neuron 17: 181–190

Tsachaki M, Ghiso J, Rostagno A, Efthimiopoulos S (2008) BRI2 homodimerizes with the involvement of intermolecular disulfide bonds. Neurobiol Aging, in press

Vassalli P (1992) The pathophysiology of tumor necrosis factors. Ann Rev Immunol 10: 411–452

Vetrivel KS, Zhang X, Meckler X, Cheng H, Lee S, Gong P, Lopes KO, Chen Y, Iwata N, Yin KJ, Lee JM, Parent AT, Saido TC, Li YM, Sisodia SS, Thinakaran G (2008) Evidence that CD147 modulation of Abeta levels is mediated by extracellular degradation of secreted Abeta. J Biol Chem, in press

Wang J, Beher D, Nyborg AC, Shearman MS, Golde TE, Goate A (2006a) C-terminal PAL motif of presenilin and presenilin homologues required for normal active site conformation. J Neurochem 96: 218–227

Wang Y, Zhang Y, Ha Y (2006b) Crystal structure of a rhomboid family intramembrane protease. Nature 444: 179–180

Weggen S, Eriksen JL, Das P, Sagi SA, Wang R, Pietrzik CU, Findlay KA, Smith TE, Murphy MP, Bulter T, Kang DE, Marquez-Sterling N, Golde TE, Koo EH (2001) A subset of NSAIDs lower amyloidogenic Abeta42 independently of cyclooxygenase activity. Nature 414: 212–216

Weihofen A, Martoglio B (2003) Intramembrane-cleaving proteases: controlled liberation of functional proteins and peptides from membranes. Trends Cell Biol 13: 71–78

Weihofen A, Binns K, Lemberg MK, Ashman K, Martoglio B (2002) Identification of signal peptide peptidase, a presenilin-type aspartic protease. Science 296: 2215–2218

Weihofen A, Lemberg MK, Friedmann E, Rueeger H, Schmitz A, Paganetti P, Rovelli G, Martoglio B (2003) Targeting presenilin-type aspartic protease signal peptide peptidase with gamma-secretase inhibitors. J Biol Chem 278: 16528–16533

Wolfe MS, Kopan R (2004) Intramembrane proteolysis: theme and variations. Science 305: 1119–1123

Wolfe MS, Xia W, Ostaszewski BL, Diehl TS, Kimberly WT, Selkoe DJ (1999) Two transmembrane aspartates in presenilin-1 required for presenilin endoproteolysis and γ-secretase activity. Nature 398: 513–517

Wu Z, Yan N, Feng L, Oberstein A, Yan H, Baker RP, Gu L, Jeffrey PD, Urban S, Shi Y (2006) Structural analysis of a rhomboid family intramembrane protease reveals a gating mechanism for substrate entry. Nature Struct Mol Biol 13: 1084–1091

Yamasaki A, Eimer S, Okochi M, Smialowska A, Kaether C, Baumeister R, Haass C, Steiner H (2006) The GxGD motif of presenilin contributes to catalytic function and substrate identification of gamma-secretase. J Neurosci 26: 3821–3828

Ye J, Rawson RB, Komuro R, Chen X, Dave UP, Prywes R, Brown MS, Goldstein JL (2000) ER stress induces cleavage of membrane-bound ATF6 by the same proteases that process SREBPs. Mol Cell 6: 1355–1364.

Yu G, Nishimura M, Arawaka S, Levitan D, Zhang L, Tandon A, Song YQ, Rogaeva E, Chen F, Kawarai T, Supala A, Levesque L, Yu H, Yang DS, Holmes E, Milman P, Liang Y, Zhang DM, Xu DH, Sato C, Rogaev E, Smith M, Janus C, Zhang Y, Aebersold R, Farrer LS, Sorbi S, Bruni A, Fraser P, St George-Hyslop P (2000) Nicastrin modulates presenilin-mediated notch/glp-1 signal transduction and βAPP processing. Nature 407: 48–54.

Zhou S, Zhou H, Walian PJ, Jap BK (2005) CD147 is a regulatory subunit of the gamma-secretase complex in Alzheimer's disease amyloid beta-peptide production. Proc Natl Acad Sci USA102: 7499–7504

Axonal Transport and Neurodegenerative Disease

Erika L. F. Holzbaur

Abstract Active intracellular transport is required to maintain the extended cellular processes of neurons. Long-distance transport along the axon is mediated by molecular motor proteins moving along the microtubule cytoskeleton. Members of the *kinesin* family drive anterograde transport, from cell body to cell periphery. Traffic back to the cell body is driven by the microtubule motor *cytoplasmic dynein* and its activator *dynactin*. Recent progress has provided insights into the mechanisms of motor protein function in axonal transport and the role of the microtubule-associated protein *tau* in regulating the spatial and temporal dynamics of microtubule motors. The dependence of the neuron on active axonal transport suggests that defects in the process might be causally linked to neurodegeneration. In particular, dynein-mediated transport is required to target old/misfolded/aggregated proteins for degradation, as well as to mediate the trophic factor signaling required to maintain a healthy neuron. The identification of disease-causing mutations in the retrograde motor complex has provided support for this hypothesis. Specifically, multiple mutations in the dynein heavy chain gene have been shown to cause neurodegeneration in the mouse, and a point mutation in the $p150^{Glued}$ subunit of dynactin has been identified as a cause of motor neuron degeneration in a human cohort. More generally, defects in axonal transport have been observed in a range of neurodegenerative diseases. We now propose a model in which both decreased efficiency of retrograde transport and alterations in retrograde signaling contribute to neurodegenerative disease.

E.L.F. Holzbaur
University of Pennsylvania School of Medicine, D400 Richards Building, 3700 Hamilton Walk, Philadelphia, PA 19104-6085, USA, E-mail: holzbaur@mail.med.upenn.edu

P. St. George-Hyslop et al. (eds.) *Intracellular Traffic and Neurodegenerative Disorders*, Research and Perspectives in Alzheimer's Disease,

1 Introduction

Neurons are highly polarized cells characterized by axons that can extend over long distances. The length of these axons makes neurons particularly dependent on active intracellular transport. Transport within the neuron is dependent on molecular motors that move cargo along the cellular cytoskeleton along both microtubules and actin filaments. Long-distance transport occurs primarily along microtubules, whereas shorter-distance movements are actin-based. Both microtubule-based and actin-based motors are required for normal neuronal function, and defects in these proteins have been linked to a range of neurodegenerative diseases. Defects in motor protein function may result directly from specific mutations in the genes encoding the motors, their activators or adaptors, or may be a downstream consequence of changes in the cellular environment, leading to altered regulation of motor activity. Here, our current understanding of the motors that drive axonal transport and the links between defects in these motors and neurodegenerative disease is reviewed.

2 Axonal Transport

Microtubules form a polarized network extending radially outward from the cell center. In axons, microtubules are organized with uniform polarity; the more dynamic plus ends of microtubules are oriented outward, with the less dynamic minus ends of the microtubule oriented inward toward the cell center. Microtubule organization in dendrites is more complex; microtubules are oriented with a mixed polarity. The reason for this increased complexity in dendrites is unclear, but one possibility is that this distinct organization may provide spatial cues governing axon/dendrite specification and maintenance.

Motor proteins move primarily unidirectionally along cytoskeletal tracks, so that the polarity of the microtubule determines the direction of transport. Transport outward from the cell center is driven by microtubule plus end-directed motors that are members of the *kinesin* superfamily. Kinesins share a common motor domain coupled to more specialized cargo-binding domains. Multiple kinesins drive *anterograde transport*; there is evidence for functional specification of motor for cargo (reviewed in Hirokawa and Takemura 2005). Kinesin-driven transport is required to supply the distal axon and neuromuscular junction with newly synthesized material, such as the proteins and lipids required for synaptic vesicle assembly and release.

Transport inward, from the axon terminal back to the cell center, is driven by the microtubule motor *cytoplasmic dynein*. Dynein is a large, multi-subunit protein complex with a motor domain within the dynein heavy chain that is homologous to the AAA family of ATPases. The active motor complex includes intermediate, light intermediate, and light chains that mediate interactions with various adaptors, effectors, and cargo molecules. One of these interacting proteins is *dynactin*, a required activator for most of the cellular functions of cytoplasmic dynein. Dynactin is also a large, multi-subunit protein complex that includes the p150^{Glued}, dynamitin, and Arp1 subunits.

Retrograde transport driven by the cytoplasmic dynein/dynactin complex is required for cellular maintenance. Dynein and dynactin mediate the transport of misfolded and aggregated proteins to the cell center for efficient degradation via the lysosomal and autophagic pathways (Jordens et al. 2001; Ravikumar et al. 2005). However, retrograde transport is also a critical mediator of intracellular signaling pathways, such as neurotrophic factor signaling (Heerssen et al. 2004) and injury response signaling pathways (Perlson et al. 2005). Increasing evidence indicates that appropriate intracellular trafficking modulates key signaling pathways (Taub et al. 2007), suggesting that there are both spatial and temporal aspects to signal transduction cascades in the cell.

Actin-based motors of the myosin superfamily also contribute to intracellular motility in the neuron, but this motility is primarily for short-distance transport in the actin-rich networks near the cell cortex. For example, myosin V appears to play critical roles in short-range transport at dendritic spines and axon terminals (reviewed in Langford 2002). More research is required to investigate the regulated switching of cargo motility from microtubules to actin and vice versa, as this process is also likely to be critical for normal neuronal function.

3 Transport in the Complex Cellular Environment

Motor proteins such as dynein and kinesin exhibit robust motility in the cell, as demonstrated by live cell imaging of labeled cargo. These motors also exhibit robust motility in vitro in assays using purified or recombinant motor proteins moving along isolated microtubules. The assays have shown that the two major axonal motors have fundamentally different properties. Kinesin moves in a hand-over-hand mechanism in 8-nm steps along the microtubule, resulting in highly processive movement in a linear pathway that closely follows the protofilament substructure of the microtubule (reviewed in Vale 2003). In contrast, single molecules of cytoplasmic dynein can take steps of variable size as they move along (Mallik et al. 2004) and across the microtubule surface (Wang et al. 1995); single molecules of dynein also take a significant number of backward (plus end-directed) steps (Ross et al. 2006). However, some of this variability is damped down when multiple dynein motors are engaged in moving a single cargo (Mallik et al. 2005).

While these in vitro assays generally model the velocity of cargo transport observed in vivo, there are often significant differences in factors such as run length and directional switching. To more fully understand motor function in the cell, it is important to more closely model intracellular conditions. For example, microtubules in the cell are not smooth and empty highways allowing for optimum motor efficiency. Instead, microtubules are decorated extensively with binding proteins known as *microtubule-associated proteins (MAPs)*. Neuronal MAPs, such as MAP2 and tau, are found in a polarized distribution in the cell; while MAP2 is localized to the somatodendritic compartment, tau is spatially restricted to axons.

The effects on motor function of MAP binding to the microtubule can be assayed directly, using high-resolution assays. In assays using kinesin bound to beads, Vershinin et al. (2006) observed that movement of the beads along microtubules was significantly inhibited by the presence of tau. Specifically, motor binding and run length were both inhibited. More recently, Dixit et al. (2008) compared the effects of tau on the movement of individual fluorescently labeled kinesin and dynein molecules. Kinesin is significantly inhibited by tau at concentrations similar to expression levels in vivo. In contrast, inhibition of dynein motility by tau was only observed at much higher concentrations and is, therefore, unlikely to be physiologically relevant. Fluorescent labeling of both kinesin and tau allows for the visualization of individual encounters with single molecule resolution. These studies have shown that kinesin is likely to either pause or dissociate from the microtubules as the motor encounters tau bound along its track (Dixit et al. 2008). In contrast, dynein is more likely to move smoothly along, much less affected by the presence of tau. Intriguingly, these in vitro observations are consistent with the cellular observations of Mandelkow et al. (2003), demonstrating that the misregulation of tau preferentially affects fast anterograde transport, driven by kinesin.

Together, these data fit a model in which microtubule-associated proteins such as tau may locally control motor function. For example, tau is not uniformly distributed along the axon (Kempf et al. 1996; Black et al. 1996). Lower levels of tau at the proximal end of the axon would allow efficient initiation of kinesin transport outward along the process. As tau levels increase distally, this would lead to a higher likelihood of kinesin detachment from the microtubule, potentially allowing for effective distribution of kinesin-bound cargos along the process. In contrast, cytoplasmic dynein initiates transport distally at the axon tip, where tau levels may be highest. Thus, it is important that dynein can effectively negotiate transport along a microtubule despite a high local concentration of MAPs. This model allows for cell-specific regulation, as multiple tau isoforms are expressed in the neuron and these isoforms show pronounced differences in their effects on motor function (Vershinin et al. 2006; Dixit et al. 2008). Further, in some neurodegenerative diseases such as Alzheimer's, there is altered expression and phosphorylation of tau as well as accumulation of the protein in the somatodendritic compartment (Khatoon et al. 1992; Braak et al. 1994); this mislocalization of tau could in turn lead to further degeneration through deleterious effects on microtubule-based transport.

It should be noted that Yuan et al. (2008) did not see significant effects of tau deletion or overexpression in vivo on rates of anterograde transport, so the consequences of tau mis-localization, mis-expression, or mis-regulation require further analysis. Still it is clear that building increasingly accurate in vitro models of intracellular transport will provide significant further insight into the mechanisms involved in this essential process.

4 Defects in Axonal Transport Lead to Neurodegeneration

The essential nature of active transport in a cell with processes that can extend up to a meter in length makes axonal transport a potential point of vulnerability, an Achilles heel for the neuron. Therefore, it has been hypothesized that defects in axonal transport could be sufficient to induce distal axonal degeneration and cell death. There is accumulating evidence that mutations in motor or cytoskeletal proteins can cause neurodegeneration (reviewed in Chevalier-Larsen and Holzbaur 2006). Here, we will focus on recent progress in our understanding of the retrograde transport pathway and its role in neurodegeneration. We will consider degeneration caused directly by mutations in either dynein or dynactin, as well as perturbations of axonal transport that may arise more indirectly, through the expression of toxic proteins such as mutant forms of *superoxide dismutase 1 (SOD1)*, as seen in the inherited motor neuron disease *amyotrophic lateral sclerosis (ALS)*.

The hypothesis that disruption of retrograde axonal transport is sufficient to induce motor neuron degeneration was first tested in a transgenic mouse with a targeted inhibition of dynein/dynactin function. Overexpression of the dynactin subunit dynamitin in motor neurons led to the disruption of dynein/dynactin function and inhibition of retrograde transport (LaMonte et al. 2002). $\mathrm{Tg}^{\mathrm{dynamitin}}$ mice display late-onset, slowly progressive degeneration of motor neurons. There is a preferential loss of large-caliber motor neurons leading to muscle atrophy.

Genetic evidence linking defects in dynein function to neurodegenerative disease was initially provided by the analysis of two lines of mice produced by N-ethyl-N-nitrosourea (ENU) mutagenesis. Both the *Loa* and *Cra1* lines encode point mutations in the gene encoding the heavy chain of cytoplasmic dynein. Neither the F580Y Loa mutation nor the Y1055C Cra1 mutation maps within the dynein motor domain. Instead, both of the mutations localize to regions of the dynein heavy chain predicted to mediate inter-subunit interactions within the complex, either dimerization of the two heavy chains or association with the dynein intermediate chain. Both lines exhibit late-onset, slowly progressive neuronal degeneration in heterozygous animals; the mutations are lethal when homozygous (Hafezparast et al. 2003).

More recently, the radiation-induced *Sprawling (Swl)* mouse has been determined to carry a nine base-pair deletion in the gene encoding cytoplasmic dynein heavy chain, leading to the in-frame substitution of alanine for amino acid residues GIVT at positions 1040–1043 (Chen et al. 2007). While this mutation in the dynein heavy chain maps relatively closely to the Cra1 mutation discussed above, mice carrying the Swl mutation display an early-onset sensory neuropathy and defective proprioception (Chen et al. 2007). These investigators demonstrated a similar proprioception defect in the Loa mouse, further supporting a key role for dynein in sensory neurons.

As a single major form of cytoplasmic dynein drives retrograde axonal transport in all types of neurons, the observation that mutations in dynein in the mouse lead to both sensory neuropathy and motor neuron degeneration is not unexpected. However, these observations suggest that some types of neurons may be more sensitive to alterations in dynein function than others. There are several possibilities in this regard. First, a subtle inhibition of dynein function, as seen for the Loa mutation (Kieran et al. 2005), may be more likely to deleteriously affect neurons with the largest or longest cellular projections. Second, a subtle mutation in dynein function may be more likely to deleteriously affect neurons with the greatest reliance on trophic factor support from the cell periphery, leading to either developmental or degenerative defects. And third, it is possible that the individual mutations identified may preferentially disrupt specific motor-cargo interactions. As the nature of the dynein-cargo interaction is not yet well understood, further analysis of these mutations and their effects on dynein function may provide further insight into the cargo selectivity of axonal transport.

The simplest hypothesis to explain the neurodegeneration observed in the $Tg^{dynamitin}$, Loa, Cra1, and Swl mice is that the targeted disruption or mutations inhibit dynein function, resulting in a perturbation of axonal transport. The effects of the Loa mutation on intracellular transport have been investigated in cellular assays (Kieran et al. 2005). Transport of a retrograde tracer, a fragment of tetanus toxin, was not significantly affected in embryonic motor neurons cultured from heterozygous Loa/+ mice, suggesting that the impairment is relatively subtle in this assay. Embryonic motor neurons cultured from homozygous Loa mice show a more significant defect in retrograde axonal transport in this assay, with a significant reduction in the frequency of high speed transport of the tetanus toxin tracer and a marked increase in pausing (Hafezparast et al. 2002). In vivo measures of retrograde transport in the $Tg^{dynamitin}$ mouse have shown a similar slowing of transport, but not a complete block (LaMonte et al. 2002).

Since dynein and dynactin are required for cell division and throughout development, loss of either protein is lethal early in embryogenesis (Harada et al. 1998; Laird et al. 2007). Thus, animals with only relatively subtle mutations, or with mutations that affect only a specific cellular function, are likely to survive. As discussed above, the Loa, Cra1, and Swl mutations in dynein heavy chain in the mouse all map to the same protein domain, so it will be of interest to test for specific effects of these mutations on neuronal functions, such as association with neuro-specific cargo. However, the Loa mutation does subtly affect dynein/dynactin function in nonneuronal cells (Hafezparast et al. 2003), indicating that while the observed phenotype of this mutation is neuronal, the disruption of function is not strictly neuron-specific. Therefore, it is important to consider possible effects of these mutations on other functions of dynein in the cell, such as trafficking in either the synthetic (ER-to-Golgi) or degradative (lysosomal and autophagic) pathways, as well as developmental pathways, such as neuronal migration and branching.

5 Mutant Dynactin in Motor Neuron Disease

Clear demonstration of a link between the dynein/dynactin complex and motor neuron disease in humans came from the identification of a mutation in the *DCTN1 gene* encoding the $p150^{Glued}$ subunit of dynactin as the cause of an inherited form of motor neuron disease (Puls et al. 2003). Patients with a G59S mutation in the DCTN1 gene display an autosomal dominant, slowly progressive form of inherited motor neuron disease. Onset in early adulthood is marked by vocal fold paralysis leading to breathing difficulties, as well as progressive weakness of the muscles in the face and hands, with distal limb atrophy developing later (Puls et al. 2003, 2005).

Dynactin is a required activator for most dynein functions in the cell, enhancing the processivity of the motor and linking the motor to some intracellular cargos (reviewed in Schroer 2004). The G59S mutation in the $p150^{Glued}$ subunit of dynactin occurs in a critical domain of the polypeptide, the highly conserved N-terminal CAP-Gly motif that mediates binding to microtubules (Waterman-Storer et al. 1995). Biochemical and cellular studies demonstrate that the mutation inhibits dynactin, resulting in some loss-of-function; binding of $p150^{Glued}$ to microtubules and to the microtubule plus end tracking protein EB1 is inhibited (Puls et al. 2003; Levy et al. 2006).

However, the key defect induced by the G59S mutation may be the distortion of folding of the $p150^{Glued}$ polypeptide, inducing aggregation in a dominant negative gain-of-function. Modeling studies suggest that the increased size of the serine side chain in the mutant protein results in steric hindrance and distortion of folding (Puls et al. 2003). At the cellular level, over-expression of the G59S mutation results in the formation of intracellular protein aggregates (Levy et al. 2006). The aggregates observed in cellular model systems closely resemble those observed in spinal cord sections from a patient expressing the G59S mutation, which are immunopositive for both dynactin and dynein (Puls et al. 2005). Together, these results support the hypothesis that the G59S mutation in $p150^{Glued}$ leads to misfolding of the polypeptide, resulting in both loss-of-function and toxic gain-of-function.

This hypothesis has now been tested in a series of mouse models expressing the G59S mutation in $p150^{Glued}$. These include two transgenic mouse models expressing low levels of the G59S polypeptide in motor neurons driven by the Thy1 promoter (Chevalier-Larsen et al. 2008; Laird et al. 2008) and a knock-in model developed by Lai et al. (2007). Both the heterozygous knock-in mice and the Tg^{G59S} model from Chevalier-Larsen et al. (2008) display a very subtle phenotype limited to mild gait abnormalities for the knock-in model and progressive weakness in grip strength assays for the transgenic model. While both models show little to no loss of motor neurons, both show clear evidence for distal degeneration, most apparent in the destabilization of the neuromuscular junction. More proximal changes are also observed; the Tg^{G59S} model exhibits a significant proliferation of tertiary lysosomes and lipofuscin granules not seen in the Loa mutant model described above (Chevalier-Larsen et al. 2008). Surprisingly, analysis of axonal transport in vivo in the Tg^{G59S} model shows that expression of the mutant transgene does not significantly affect retrograde transport (Chevalier-Larsen et al. 2008).

In contrast, a second transgenic model, developed by Laird et al. (2008), displays much more significant neurodegeneration, leading to early death. This more rapidly progressive model displays motor neuron loss and axonal swelling, as well as distal degeneration. In this model, there is evidence for the accumulation of the mutant polypeptide in prominent intracellular inclusions, some of which co-localize with autophagosome markers. The reason for the significant disparity in phenotype observed in the two transgenic models is unclear but may be a consequence of relative expression levels. The course of disease in human patients does not closely correspond to the rapidly progressive phenotype observed in the model from Laird et al. (2008), but these mice do model aggregate formation that in some respects is similar to that seen in immunohistological analysis of affected human spinal cords.

The studies of Lai et al. (2007) demonstrate that the DCTN1 gene encoding p150^{Glued} is haplo-sufficient, so while the G59S mutation adversely affects dynactin activity, loss-of-function is not the major problem. Instead, misfolding of the mutant polypeptide, leading to aggregation and/or altered intracellular trafficking to lysosomes and autophagosomes, is seen in the transgenic models. These observations are consistent with the dominant phenotype of the G59S mutation in the affected kindred (Puls et al. 2003). Further, the pronounced distal degeneration and destabilization of the neuromuscular junction seen in all three models support a critical role for dynein in multiple cellular pathways. It is particularly striking that both distal degeneration and altered intracellular trafficking are seen in a mouse model without a significant defect in retrograde axonal transport (Chevalier-Larsen et al. 2008). Therefore, these studies suggest that perturbation of dynein and dynactin in the cell may have multiple deleterious effects, consistent with the pleiotropic cellular role of this motor complex.

6 Do Defects in Axonal Transport Contribute to Pathogenesis in ALS?

While it is now clear that mutations in dynein or dynactin can cause neuronal degeneration directly, it is also possible that alterations in dynein/dynactin function that are caused more indirectly may contribute to disease pathogenesis during motor neuron degeneration. For example, either mutations in interacting proteins or changes in the cellular environment may significantly affect cytoskeletal integrity or motor function, leading in turn to further degeneration. Or, the accumulation of toxic protein aggregates might impede efficient transport along the axon, leading to distal axonal degeneration. More broadly, the key role of retrograde transport in pathways such as neurotrophic factor signaling suggests that inhibition of transport may lead to cell death through the perturbation of signal transduction pathways or the activation of stress/death pathways (Perlson and Holzbaur, preliminary data). The balance between cell survival and cell stress pathways may therefore be directly affected by alterations in axonal transport in the neuron.

Specifically, defects in axonal transport have been proposed to contribute to the death of motor neurons observed in ALS. ALS is a fatal neurodegenerative disease characterized by the loss of motor neurons, leading to muscle atrophy. Most cases of ALS are sporadic, with no known cause. Approximately 10% of ALS cases are genetic, with mutations in the superoxide dismutase 1 (SOD1) gene as the most common identified cause. These mutations appear to induce a toxic gain-of-function, likely involving protein misfolding and aggregation (reviewed in Bruijn et al. 2004).

Initial studies in mouse models of familial ALS expressing mutant SOD1 (mSOD1), which exhibit rapid and dramatic loss of motor neurons (Gurney et al. 1994), led to the suggestion that axonal aggregates of mutant SOD1 might physically block transport. Inhibition of slow axonal transport was observed, including perturbations in the transport of neurofilaments and tubulin, observed just prior to significant degeneration and loss of motor neurons (Zhang et al. 1997; Williamson and Cleveland 1999). Fast anterograde transport was not generally inhibited prior to disease onset.

In contrast, defects in retrograde transport occur much earlier in the disease process in the mSOD1 model. In studies in which a retrograde neurotracer was injected into muscle and transport to the cell bodies of motor neurons was assayed, significant inhibition of transport was observed in mSOD1 mice (Ligon et al. 2005). The observed inhibition initiated at a very early point, well prior to the onset of clinical disease. Studies on axonal transport in embryonic motor neurons cultured from the mSOD1 mouse also showed a significant slowing of retrograde transport. Live cell analysis of the retrograde motility of a fluorescently labeled fragment of tetanus toxin showed a shift in the speed profile, indicative of an increased frequency of pauses and oscillatory movements, as compared to the more efficient transport observed in motor neurons from wild type mice (Kieran et al. 2005).

What has yet to be determined is how the expression of mutant SOD1 can disrupt dynein function. One possibility is that the disruption is direct. Dynein co-localizes with SOD1 aggregates in neurons expressing mutant SOD1 (Ligon et al. 2005); these aggregates may deplete the pool of active motors along the axon. A biochemical interaction between SOD1 and dynein has also been described (Zhang et al. 2007), although the affinity of this interaction is relatively low (Perlson and Holzbaur, preliminary data). Evidence to date does not support an early hypothesis that aggregates of misfolded protein result in a physical block of transport. Instead, the mechanism may be indirect, due to the activation of stress response pathways caused by the accumulation of misfolded SOD1 protein. This activation appears to alter the regulatory balance in the axon, leading to a mis-regulation of the motors driving axonal transport (Perlson and Holzbaur, preliminary data).

It is also possible that there is a significant change in the type of cellular compartment undergoing transport during neurodegeneration (Nixon 2005). Intriguing data supporting this possibility come from a recent study from the Mobley lab, investigating retrograde transport in a mouse model of Down's syndrome. Salehi et al. (2006) noted both the decreased transport of NGF and a corresponding enlargement of early endosomes in neurons from affected mice. An apparent stalling of early endosomes along the axon was observed. It will be interesting to determine whether a similar effect is seen in other models, such as the mSOD1 mouse, thus indicating a common pathway involving alterations in the retrograde transport compartment.

While retrograde axonal transport represents one key role for dynein and dynactin in the neuron, an additional cellular function of dynein and dynactin may also be critical to the health and function of motor neurons. Studies in Drosophila have shown that dynactin is essential to maintain the integrity of *neuromuscular junctions* (NMJs; Eaton et al., 2002). Disruption of dynactin by either RNAi or expression of a mutant form of p150^{Glued} leads to local disruption of the microtubule cytoskeleton in presynaptic motor neuron nerve terminals, followed by synapse retraction.

This observation is particularly intriguing, given the observations of NMJ degeneration in mice expressing the G59S mutant form of dynactin (Lai et al. 2007; Laird et al. 2008; Chevalier-Larsen et al. 2008), as well as the early and progressive loss of NMJs in the mSOD1 model for familial ALS (Fischer et al. 2004). Further work will be required to determine if this axonal die-back is a downstream consequence of defects in dynein/dynactin-mediated trafficking or instead is due to disruption of an additional function for dynactin in maintaining junctional integrity.

7 Conclusions

Increasingly, defects in transport, trafficking, and the cytoskeleton are being linked to the development of neurodegenerative disease. As noted above, the identification of the G59S mutation in the DCTN1 gene in humans and the Loa, Cra1, and Swl mutations in the Dync1h1 gene in mouse has provided strong support for the hypothesis that active axonal transport is critical to maintain the health of neurons. Additional mutations in dynactin have been identified in patients with either familial or sporadic ALS (Munch et al. 2004, 2005), although these have not yet been causally linked to disease.

More generally, mutations have been identified in familial ALS patients in proteins involved in vesicular trafficking, including alsin and VAMP-associated protein B (Hadano et al. 2001; Yang et al. 2001; Nishimura et al., 2004). In other neurodegenerative diseases, mutations in motors from the kinesin superfamily have been identified as the cause of Charcot-Marie-Tooth Disease Type 2A (Zhao et al. 2001) and of congenital fibrosis of the extraocular muscles type 1 (CFEOM1; Yamada et al. 2003). Mutations in a kinesin gene (SPG10) also cause one form of hereditary spastic paraplegia (Reid et al. 2002).

Further, mutations in other proteins linked directly to neurodegeneration, such as Huntingtin or amyloid precursor protein (APP), also affect axonal transport either directly or indirectly. Together, these data suggest that neurons are absolutely dependent on axonal transport and are uniquely vulnerable to defects in the motors and other cytoskeletal proteins involved in this transport. A better understanding of the underlying cell biology of motor neurons will therefore be required to fully understand the role of transport defects in the pathogenesis of neurodegenerative disease.

References

Black MM, Slaughter T, Moshiach S, Obrocka M, Fischer I (1996) Tau is enriched on dynamic microtubules in the distal region of growing axons. J Neurosci 16:3601–3619.

Boillee S, Vande Velde C, Cleveland DW (2006) ALS: a disease of motor neurons and their nonneuronal neighbors. Neuron 52:39–59.

Braak E, Braak H, Mandelkow EM (1994) A sequence of cytoskeleton changes related to the formation of neurofibrillary tangles and neuropil threads. Acta Neuropathol 87:554–567.

Bruijn LI, Miller TM, Cleveland DW (2004) Unraveling the mechanisms involved in motor neuron degeneration in ALS. Annu Rev Neurosci 27:723–749.

Chen XJ, Levedakou EN, Millen KJ, Wollmann RL, Soliven B, Popko B (2007) Proprioceptive sensory neuropathy in mice with a mutation in the cytoplasmic dynein heavy chain 1 gene. J Neurosci 27:14515–14524.

Chevalier-Larsen E, Holzbaur EL (2006) Axonal transport and neurodegenerative disease. Biochim Biophys Acta 1762:1094–1108.

Chevalier-Larsen ES, Wallace KE, Pennise CR, Holzbaur EL (2008) Lysosomal proliferation and distal degeneration in motor neurons expressing the G59S mutation in the p150Glued subunit of dynactin. Human Mol Genet, in press

Dixit R, Ross JL, Goldman YE, Holzbaur EL (2008) Differential regulation of dynein and kinesin motor proteins by tau. Science 319:1086–1089.

Eaton BA, Fetter RD, Davis GW (2002) Dynactin is necessary for synapse stabilization. Neuron 34:729–741.

Fischer LR, Culver DG, Tennant P, Davis AA, Wang M, Castellano-Sanchez A, Khan J, Polak MA, Glass JD (2004) Amyotrophic lateral sclerosis is a distal axonopathy: evidence in mice and man. Exp Neurol 185:232–240.

Gurney ME, Pu H, Chiu AY, Dal Canto MC, Polchow CY, Alexander DD, Caliendo J, Hentati A, Kwon YW, Deng HX, et al. (1994) Motor neuron degeneration in mice that express a human Cu,Zn superoxide dismutase mutation. Science 264:1772–1775.

Hadano S, Hand CK, Osuga H, Yanagisawa Y, Otomo A, Devon RS, Miyamoto N, Showguchi-Miyata J, Okada Y, Singaraja R, Figlewicz DA, Kwiatkowski T, Hosler BA, Sagie T, Skaug J, Nasir J, Brown RH, Jr., Scherer SW, Rouleau GA, Hayden MR, Ikeda JE (2001) A gene encoding a putative GTPase regulator is mutated in familial amyotrophic lateral sclerosis 2. Nature Genet 29:166–173.

Hafezparast M, Klocke R, Ruhrberg C, Marquardt A, Ahmad-Annuar A, Bowen S, Lalli G, Witherden AS, Hummerich H, Nicholson S, Morgan PJ, Oozageer R, Priestley JV, Averill S, King VR, Ball S, Peters J, Toda T, Yamamoto A, Hiraoka Y, Augustin M, Korthaus D, Wattler S, Wabnitz P, Dickneite C, Lampel S, Boehme F, Peraus G, Popp A, Rudelius M, Schlegel J, Fuchs H, Hrabe de Angelis M, Schiavo G, Shima DT, Russ AP, Stumm G, Martin JE, Fisher EM (2003) Mutations in dynein link motor neuron degeneration to defects in retrograde transport. Science 300:808–812.

Harada A, Takei Y, Kanai Y, Tanaka Y, Nonaka S, Hirokawa N (1998) Golgi vesiculation and lysosome dispersion in cells lacking cytoplasmic dynein. J Cell Biol 141:51–59.

Heerssen HM, Pazyra MF, Segal RA (2004) Dynein motors transport activated Trks to promote survival of target-dependent neurons. Nature Neurosci 7:596–604.

Hirokawa N, Takemura R (2005) Molecular motors and mechanisms of directional transport in neurons. Nature Rev Neurosci 6:201–214.

Jordens I, Fernandez-Borja M, Marsman M, Dusseljee S, Janssen L, Calafat J, Janssen H, Wubbolts R, Neefjes J (2001) The Rab7 effector protein RILP controls lysosomal transport by inducing the recruitment of dynein-dynactin motors. Curr Biol 11:1680–1685.

Kempf M, Clement A, Faissner A, Lee G, Brandt R (1996) Tau binds to the distal axon early in development of polarity in a microtubule- and microfilament-dependent manner. J Neurosci 16:5583–5592.

Khatoon S, Grundke-Iqbal I, Iqbal K (1992) Brain levels of microtubule-associated protein tau are elevated in Alzheimer's disease: a radioimmuno-slot-blot assay for nanograms of the protein. J Neurochem 59:750–753.

Kieran D, Hafezparast M, Bohnert S, Dick JR, Martin J, Schiavo G, Fisher EM, Greensmith L (2005) A mutation in dynein rescues axonal transport defects and extends the life span of ALS mice. J Cell Biol 169:561–567.

Lai C, Lin X, Chandran J, Shim H, Yang WJ, Cai H (2007) The G59S mutation in p150(glued) causes dysfunction of dynactin in mice. J Neurosci 27:13982–13990.

Laird FM, Farah MH, Ackerley S, Hoke A, Maragakis N, Rothstein JD, Griffin J, Price DL, Martin LJ, Wong PC (2008) Motor neuron disease occurring in a mutant dynactin mouse model is characterized by defects in vesicular trafficking. J Neurosci 28:1997–2005.

LaMonte BH, Wallace KE, Holloway BA, Shelly SS, Ascano J, Tokito M, Van Winkle T, Howland DS, Holzbaur EL (2002) Disruption of dynein/dynactin inhibits axonal transport in motor neurons causing late-onset progressive degeneration. Neuron 34:715–727.

Langford GM (2002) Myosin-V, a versatile motor for short-range vesicle transport. Traffic 3: 859–865.

Levy JR, Holzbaur EL (2006) Cytoplasmic dynein/dynactin function and dysfunction in motor neurons. Intl J Dev Neurosci 24:103–111.

Levy JR, Sumner CJ, Caviston JP, Tokito MK, Ranganathan S, Ligon LA, Wallace KE, LaMonte BH, Harmison GG, Puls I, Fischbeck KH, Holzbaur EL (2006) A motor neuron disease-associated mutation in p150Glued perturbs dynactin function and induces protein aggregation. J Cell Biol 172:733–745.

Ligon LA, LaMonte BH, Wallace KE, Weber N, Kalb RG, Holzbaur EL (2005) Mutant superoxide dismutase disrupts cytoplasmic dynein in motor neurons. Neuroreport 16:533–536.

Mallik R, Carter BC, Lex SA, King SJ, Gross SP (2004) Cytoplasmic dynein functions as a gear in response to load. Nature 427:649–652.

Mallik R, Petrov D, Lex SA, King SJ, Gross SP (2005) Building complexity: an in vitro study of cytoplasmic dynein with in vivo implications. Curr Biol 15:2075–2085.

Mandelkow EM, Stamer K, Vogel R, Thies E, Mandelkow E (2003) Clogging of axons by tau, inhibition of axonal traffic and starvation of synapses. Neurobiol Aging 24:1079–1085.

Munch C, Sedlmeier R, Meyer T, Homberg V, Sperfeld AD, Kurt A, Prudlo J, Peraus G, Hanemann CO, Stumm G, Ludolph AC (2004) Point mutations of the p150 subunit of dynactin (DCTN1) gene in ALS. Neurology 63:724–726.

Munch C, Rosenbohm A, Sperfeld AD, Uttner I, Reske S, Krause BJ, Sedlmeier R, Meyer T, Hanemann CO, Stumm G, Ludolph AC (2005) Heterozygous R1101K mutation of the DCTN1 gene in a family with ALS and FTD. Ann Neurol 58:777–780.

Nishimura AL, Mitne-Neto M, Silva HC, Richieri-Costa A, Middleton S, Cascio D, Kok F, Oliveira JR, Gillingwater T, Webb J, Skehel P, Zatz M (2004) A mutation in the vesicle-trafficking protein VAPB causes late-onset spinal muscular atrophy and amyotrophic lateral sclerosis. Am J Human Genet 75:822–831.

Nixon RA (2005) Endosome function and dysfunction in Alzheimer's disease and other neurodegenerative diseases. Neurobiol Aging 26:373–382.

Perlson E, Hanz S, Ben-Yaakov K, Segal-Ruder Y, Seger R, Fainzilber M (2005) Vimentin-dependent spatial translocation of an activated MAP kinase in injured nerve. Neuron 45: 715–726.

Puls I, Jonnakuty C, LaMonte BH, Holzbaur EL, Tokito M, Mann E, Floeter MK, Bidus K, Drayna D, Oh SJ, Brown RH, Jr., Ludlow CL, Fischbeck KH (2003) Mutant dynactin in motor neuron disease. Nat Genet 33:455–456.

Puls I, Oh SJ, Sumner CJ, Wallace KE, Floeter MK, Mann EA, Kennedy WR, Wendelschafer-Crabb G, Vortmeyer A, Powers R, Finnegan K, Holzbaur EL, Fischbeck KH, Ludlow CL (2005) Distal spinal and bulbar muscular atrophy caused by dynactin mutation. Ann Neurol 57: 687–694.

Ravikumar B, Acevedo-Arozena A, Imarisio S, Berger Z, Vacher C, O'Kane CJ, Brown SD, Rubinsztein DC (2005) Dynein mutations impair autophagic clearance of aggregate-prone proteins. Nature Genet 37:771–776.

Reid E, Kloos M, Ashley-Koch A, Hughes L, Bevan S, Svenson IK, Graham FL, Gaskell PC, Dearlove A, Pericak-Vance MA, Rubinsztein DC, Marchuk DA (2002) A kinesin heavy chain (KIF5A) mutation in hereditary spastic paraplegia (SPG10). Am J Human Genet 71:1189–1194.

Ross JL, Wallace K, Shuman H, Goldman YE, Holzbaur EL (2006) Processive bidirectional motion of dynein-dynactin complexes in vitro. Nature Cell Biol 8:562–570.

Salehi A, Delcroix JD, Belichenko PV, Zhan K, Wu C, Valletta JS, Takimoto-Kimura R, Kleschevnikov AM, Sambamurti K, Chung PP, Xia W, Villar A, Campbell WA, Kulnane LS, Nixon RA, Lamb BT, Epstein CJ, Stokin GB, Goldstein LS, Mobley WC (2006) Increased App expression in a mouse model of Down's syndrome disrupts NGF transport and causes cholinergic neuron degeneration. Neuron 51:29–42.

Schroer TA (2004) Dynactin. Annu Rev Cell Dev Biol 20:759–779.

Taub N, Teis D, Ebner HL, Hess MW, Huber LA (2007) Late Endosomal Traffic of the EGFR Ensures Spatial and Temporal Fidelity of MAPK Signaling. Mol Biol Cell 18:4698–4710.

Vale RD (2003) The molecular motor toolbox for intracellular transport. Cell 112:467–480.

Vershinin M, Carter BC, Razafsky DS, King SJ, Gross SP (2007) Multiple-motor based transport and its regulation by Tau. Proc Natl Acad Sci USA 104:87–92.

Wang Z, Khan S, Sheetz MP (1995) Single cytoplasmic dynein molecule movements: characterization and comparison with kinesin. Biophys J 69:2011–2023.

Waterman-Storer CM, Karki S, Holzbaur EL (1995) The p150Glued component of the dynactin complex binds to both microtubules and the actin-related protein centractin (Arp-1). Proc Natl Acad Sci USA 92:1634–1638.

Williamson TL, Cleveland DW (1999) Slowing of axonal transport is a very early event in the toxicity of ALS-linked SOD1 mutants to motor neurons. Nature Neurosci 2:50–56.

Yamada K, Andrews C, Chan WM, McKeown CA, Magli A, de Berardinis T, Loewenstein A, Lazar M, O'Keefe M, Letson R, London A, Ruttum M, Matsumoto N, Saito N, Morris L, Del Monte M, Johnson RH, Uyama E, Houtman WA, de Vries B, Carlow TJ, Hart BL, Krawiecki N, Shoffner J, Vogel MC, Katowitz J, Goldstein SM, Levin AV, Sener EC, Ozturk BT, Akarsu AN, Brodsky MC, Hanisch F, Cruse RP, Zubcov AA, Robb RM, Roggenkaemper P, Gottlob I, Kowal L, Battu R, Traboulsi EI, Franceschini P, Newlin A, Demer JL, Engle EC (2003) Heterozygous mutations of the kinesin KIF21A in congenital fibrosis of the extraocular muscles type 1 (CFEOM1). Nature Genet 35:318–321.

Yang Y, Hentati A, Deng HX, Dabbagh O, Sasaki T, Hirano M, Hung WY, Ouahchi K, Yan J, Azim AC, Cole N, Gascon G, Yagmour A, Ben-Hamida M, Pericak-Vance M, Hentati F, Siddique T (2001) The gene encoding alsin, a protein with three guanine-nucleotide exchange factor domains, is mutated in a form of recessive amyotrophic lateral sclerosis. Nature Genet 29: 160–165.

Yuan A, Kumar A, Peterhoff C, Duff K, Nixon RA (2008) Axonal transport rates in vivo are unaffected by tau deletion or overexpression in mice. J Neurosci 28:1682–1687.

Zhang B, Tu P, Abtahian F, Trojanowski JQ, Lee VM (1997) Neurofilaments and orthograde transport are reduced in ventral root axons of transgenic mice that express human SOD1 with a G93A mutation. J Cell Biol 139:1307–1315.

Zhang F, Strom AL, Fukada K, Lee S, Hayward LJ, Zhu H (2007) Interaction between familial amyotrophic lateral sclerosis (ALS)-linked SOD1 mutants and the dynein complex. J Biol Chem 282:16691–16699.

Zhao C, Takita J, Tanaka Y, Setou M, Nakagawa T, Takeda S, Yang HW, Terada S, Nakata T, Takei Y, Saito M, Tsuji S, Hayashi Y, Hirokawa N (2001) Charcot-Marie-Tooth disease type 2A caused by mutation in a microtubule motor KIF1Bbeta. Cell 105:587–597.

Simple Cellular Solutions to Complex Problems

Susan Lindquist(✉) and Karen L. Allendoerfer

Abstract We have developed simple cellular models of complex neurodegenerative diseases by over-expressing disease-associated human proteins in yeast. By combining the unique power of yeast genetics with the highly conserved biology of protein homeostasis in all eukaryotes, we use yeast cells as "living test tubes" to investigate the mechanisms of toxicity associated with problems in protein folding, trafficking, and degradation and complement these basic studies with transcriptional analysis and high-throughput screens for toxicity modifiers.

Strong evidence links Parkinson's disease (PD) to the misfolding of alpha-synuclein (α-syn) and accumulation of cellular inclusions. In our yeast model over-expressing human α-syn, the behavior of that protein and its PD-associated mutants recapitulates many features of synuclein pathobiology, including extreme dosage sensitivity and the production of reactive oxygen species (ROS). Through a genome-wide screen we identified a set of genes, many with clear human homologs, that robustly modify the toxic effects of α-syn over-expression. The largest class of hits includes proteins functioning in vesicle trafficking, including Ypt1p, a Rab GTPase, that is also associated with the cytoplasmic inclusions. The earliest cellular defect in yeast following α-syn induction is a block in vesicle trafficking from the endoplasmic reticulum (ER) to the Golgi. Elevated expression of members of its family of mammalian homologs protects against dopaminergic (DA) neuron loss in whole-animal models of PD and in cultures of rat midbrain, demonstrating the relevance of results obtained in yeast to mammalian neurons. In addition to genes that play a role in vesicular trafficking, we also found hits in the categories of protein phosphorylation, nitrosative stress, and the target of rapamycin (TOR) pathway. Furthermore, we found that one of our genetic toxicity suppressors, YOR291W, is the yeast homolog of the human PD-associated gene, PARK9. We also screened a library of small molecules for chemical toxicity suppressors. Strikingly, the two

S. Lindquist
Whitehead Institute for Biomedical Research, and Howard Hughes Medical Institute
9 Cambridge Center, Cambridge MA 02142, USA
E-mail: Lindquist_admin@wi.mit.edu

P. St. George-Hyslop et al. (eds.) *Intracellular Traffic and Neurodegenerative Disorders*,
Research and Perspectives in Alzheimer's Disease,

most effective compounds in this chemical screen also selectively rescued cultured rat DA neurons from rotenone and from A53T mutant α-syn expression.

In our transcriptional microarray analysis, we found that the classes of genes most perturbed by overexpression of α-syn were related to mitochondrial function and the stress response, suggesting that mitochondrial defects known to play a key role in PD may be mediated or exacerbated by α-syn toxicity. We devised an integrative approach to analyzing transcriptional and genetic data in the context of known protein-protein interactions. Application of this approach to our yeast model provides a mechanistic context for both types of experimental data by uncovering pathways perturbed by α-syn over-expression and identifying drug targets whose manipulation alters cellular survival.

1 Introduction

1.1 Modeling Protein-Folding Diseases in Yeast

High-throughput screening and transcriptional analysis can provide an unprecedented genome- and proteome-wide view of cellular and molecular changes in experimental models, and this view may be extremely helpful in deciphering multifactorial neurodegenerative diseases. At present we lack the kind of deep biological understanding that leads to effective therapeutics for many of the most common and devastating of these diseases due, in part, to the absence of models that are amenable to high-throughput investigations.

The budding yeast *Saccharomyces cerevisiae* is by far the best-characterized and most readily manipulable eukaryotic cell, with a small genome and a huge variety of genetic tools that have been developed over the years. Large-scale proteomics projects have also been recently completed (Causier 2004; Kolkman et al. 2005) that identified more than 30,000 protein-protein interactions. In addition, thousands of expression profiles under diverse conditions have been done in this organism (e.g., Gasch et al. 2000). Yeast experiments are usually carried out under simple, standardized, reproducible conditions (Sherman 2002), enabling the pooling of data for enhanced rigor and computational power (e.g., Irizarry et al. 2005). When disease processes affect basic, highly conserved, eukaryotic functions such as protein folding and trafficking (protein homeostasis), yeast can provide an invaluable starting point to obtain general mechanistic insights that will be applicable to more complex organisms, including humans.

Many complex neurodegenerative diseases can be thought of as diseases of protein folding, caused by presumed alterations in an important protein's 3D structure and function. Anfinsen's principle that in a given environment a protein molecule will spontaneously assume the conformation of greatest thermodynamic stability based on its amino acid sequence (Anfinsen 1973) was transformative, but it does not address the practical problems represented by the multitude of steps and interactions proteins must undergo to achieve a functional conformation while navigating the chaotic, crowded cellular environment. A large number of highly conserved quality control mechanisms and compartmentalization processes have evolved to handle this ancient problem. These include chaperone proteins, protein remodeling factors, osmolytes, and proteolytic degradation machineries. Furthermore, yeast cells share with all eukaryotes mechanisms for the sequestration of proteins into diverse membrane-bound compartments. These have been conserved in all organisms but are particularly well conserved among eukaryotes.

This paper focuses on a set of yeast strains expressing human α-syn that were developed in our laboratory and that provide an important model of PD, Dementia with Lewy Bodies, Multiple Systems Atrophy, and other synucleinopathies. We have used this model to investigate mechanisms of α-syn cytotoxicity, to screen for novel therapeutic targets, and to obtain a comprehensive map of changes in cellular pathways that stem from the introduction and over-expression of wild-type α-syn and its disease-associated mutants. The fact that so many of the pathways observed to be perturbed in yeast by α-syn toxicity are related to previously known causes of PD establishes that α-syn's role in pathobiology is exerted through core biological pathways that have been conserved for a billion years. We realize that many vitally important aspects of human disease cannot be approached with this type of model. However, we have established that simple yeast models will provide basic insights relevant to human pathology when used as adjuncts to more complex systems.

1.2 Model of Synucleinopathies Based on Expression of Human α-Syn

To model the cell biology and pathology of synucleinopathies in a yeast cell, we fused human α-syn protein [in wildtype (WT) or mutant A53T and A30P forms] at its carboxy terminus to green fluorescent protein (GFP) and integrated this construct into the yeast genome for induction and expression (Outeiro and Lindquist 2003). These α-syn-GFP fusion proteins exhibited no evidence of proteolysis and localized similarly to untagged α-syn, as measured by immunofluorescence. The use of α-syn-GFP fusion proteins avoided the potential difficulties and inconsistencies of immunohistochemical methods seen with most membrane proteins and provided a powerful tool for studying changes in protein distribution in real time in living cells.

In engineering this model, we took advantage of two unique features of yeast biology: the ease of homologous recombination and the tightly regulated *GAL1* promoter. In different experiments we have employed cells with various numbers of integrations at the *URA3, HIS3*, or *TRP1* loci. Galactose-driven expression of the integrated α-syn construct enabled us to perform routine growth and maintenance of cells with α-syn expression repressed and to avoid any selective pressures on leaky α-syn expression that might have led to the accumulation of confounding genetic variation in our screens. After a brief period of growth in raffinose that allowed the glucose repressor to disappear, a switch from raffinose to galactose medium rapidly and synchronously induced α-syn expression in all cells throughout the culture.

2 Results

2.1 Toxicity of α-Syn is Dependent on Dosage

The dosage of α-syn is well correlated with the development of PD in patients; duplications or triplications of the human *SNCA* gene result in a clinical phenotype, as do mutations in regulatory regions (Farrer et al. 2001). In yeast, when a low level of human α-syn-GFP based on a single dose is induced, cells grow at about the same rate as cells carrying a control vector (Fig. 2 and Outeiro and Lindquist 2003). The WT α-syn-GFP protein rapidly becomes localized to the plasma membrane, with a much smaller quantity found in the cytoplasm (Fig. 1A and Outeiro and Lindquist 2003). A high concentration of α-syn at the yeast plasma membrane *in vivo* is consistent with its selective *in vitro* binding to phospholipid vesicles (Jo et al. 2000). It is also in agreement with the concentration of α-syn at the synapse in mammalian neurons. Indeed, electron microscopy of yeast cells expressing WT α-syn-GFP established that these cells do have a subtle defect in vesicle trafficking (Gitler et al. 2008). Small vesicles accumulate abnormally near the plasma membrane proximal to the peripheral ER (see Fig. 3A–C), suggesting that α-syn over-expression, even at low levels, interferes with vesicle docking or fusion (Gitler et al. 2008; see below). Yeast does not have a clear homolog for the human gene encoding α-syn, *SNCA*. However, all eukaryotic organisms have proteins with lipid membrane-binding properties that are similar to those of α-syn and that might serve similar cellular functions, although they are highly divergent in sequence. Indeed α-syn is thought to play a role in regulating normal synaptic vesicle priming before fusion (Larsen et al. 2006). α-Syn is intrinsically disordered and has a propensity to aggregate unless associated with membranes. It is likely these intrinsic biophysical properties that allow its biology and pathobiology to be conserved so well from yeast to humans.

In attempting to reconstitute the biology of α-syn in yeast, we reasoned that increasing the level of α-syn expression might exceed the quality control system of the cell and mimic an aging or disease scenario associated with α-syn toxicity, as has been suggested for human PD. Indeed, we found that integrating a second dose of α-syn-GFP into the yeast genome remarkably increased the protein's accumulation and changed its localization. After induction in the high-dose strain, α-syn protein first targeted to the membrane and formed small foci there; a few hours later, the vast majority of the α-syn appeared in large cytoplasmic inclusions, or foci (Fig. 1B). These foci were not formed simply by excess α-syn unable to find membrane binding sites; even though high-dose strains expressed more α-syn protein overall, they had much less of it present at the membrane than low-dose strains (Outeiro and Lindquist 2003; Fig. 1A,B). The disease-associated A53T α-syn mutant behaved much like WT protein at both dosages, except that small foci could be observed in some cells, even in cells expressing only a low dose of α-syn. Furthermore, A53T α-syn moved from membrane association to inclusion formation even more rapidly than WT protein (Outeiro and Lindquist 2003).

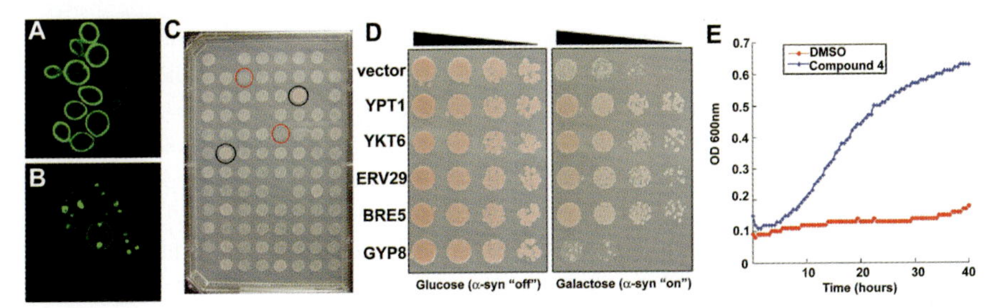

Fig. 1 Yeast Parkinson model: A discovery platform for genetic and chemical modifiers. (**A**) Expressed at low levels, α-syn fused to GFP travels with membranes through the secretory pathway, eventually concentrating at the plasma membrane, and is non-toxic. (**B**) Increasing α-syn expression causes misfolding, the formation of α-syn foci, and toxicity. (**C**) A screen of cells with intermediate toxicity (IntTox) identified plasmids that suppress toxicity (black circles) or enhance it (red circles). (**D**) Highly conserved ER-Golgi vesicle trafficking proteins potently modulate α-syn toxicity. (**E**) A compound identified in a chemical screen restores growth on solid (not shown) and liquid medium

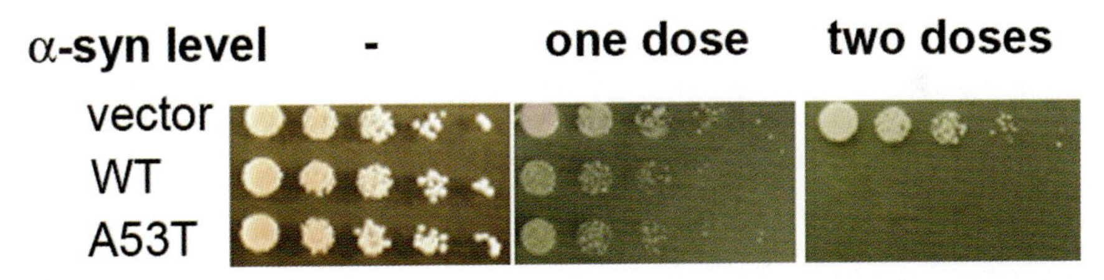

Fig. 2 α-Syn over-expression inhibits growth of yeast. One dose of α-syn, either WT or A53 mutant, integrated into the genome, had little-to-no effect on growth whereas two doses of α-syn completely inhibited growth

Taking advantage of the facts that chromosomal location can exert subtle effects on gene expression and that the extent of toxicity is dependent on α-syn expression levels, we have developed both intermediate- and high-toxicity α-syn yeast strains for different screening and transcriptional profiling purposes (Outeiro and Lindquist 2003; Cooper et al. 2006; Gitler et al. 2008Su et al., submitted for publication). We have designated these two two-dose strains HiTox (*URA3* and *TRP1* integration site; initial two-dose strain) and IntTox (*HIS3* and *TRP1* integration site; Gitler et al. 2008). As discussed below, having such precisely regulated levels of toxicity has facilitated high-throughput screening and microarray analysis, illustrating the utility of the yeast system.

To assess the extent of α-syn toxicity, we measured its effects on cell growth and viability. Cultures were plated on galactose to induce α-syn and colony growth was monitored (Fig. 2). As described above, a single dose of WT or A53T α-syn had little or no inhibitory effect on growth, whereas the doubling dose completely inhibited it (Fig. 2). To further characterize the toxicity of α-syn, we also monitored the loss of cellular viability following induction of WT α-syn in two ways, by assaying for colony-forming units remaining in the culture and by staining with the dye propidium iodide, which labels dead cells. In the HiTox strain, we observed only a few percent of cells dying at four hours post-α-syn induction, but the percentage climbed steeply thereafter, reaching ~30% cell death at six hours and ~50% at eight hours (Cooper et al. 2006 and Su et al., submitted for publication). At six hours, the IntTox strain exhibited ~10% cell death, but we have not yet characterized it as well as we have the HiTox strain (Su et al., submitted for publication).

α-Syn toxicity in our model (Outeiro and Lindquist 2003; Su et al., submitted for publication) recapitulates many aspects of α-syn toxicity in mammalian cells and PD patients, including ubiquitination of α-syn inclusions, impairment of the ubiquitin-proteasome system (Iwatsubo et al. 1996; Dauer and Przedborski 2003), A53T-mediated enhancement of toxicity, accumulation of lipid droplets (den Jager 1969; Gai et al. 2000), the production of reactive oxygen species (ROS; Giasson et al. 2000; Dauer and Przedborski 2003) and mitochondrial pathology (Ramsey and Giasson 2007). The lag time between α-syn induction and significant cell death affords an opportunity to assay the biological effects of the protein on yeast cells before irreversible toxicity confounds result.

2.2 α-Syn Over-Expression Causes a Block in Endoplasmic Reticulum-to-Golgi Vesicle Trafficking

The findings of increasingly perturbed protein homeostasis systems culminating in cell death were in keeping with observations from our collaborators in the Cooper laboratory (then at University of Missouri-Kansas City) of endoplasmic reticulum (ER) stress in the α-syn-expressing cells. They wondered if ER stress might be induced by the blockage of ER-associated degradation (ERAD) of misfolded proteins. So, they tested the degradation of two different, mutated, ERAD substrates, Sec61-2p and carboxy peptidase Y* (CPY*), in our yeast model over-expressing α-syn. Importantly, the CPY* substrate, but not Sec61-2p, requires trafficking from the ER to the Golgi prior to degradation. The degradation of Sec61-2p was unaffected in the α-syn model, but the degradation of CPY*, the substrate that requires ER-to-Golgi trafficking, was decreased. These results, along with those of our genetic screen and microarray analyses (see below), provided evidence that α-syn over-expression caused a block in ER-to-Golgi vesicle trafficking. To test this hypothesis directly, we then monitored the trafficking of two additional proteins, CPY-WT and alkaline phosphatase, both of which acquire various modifications that can be visualized on gels as they traffic from the ER through the Golgi to the vacuole. In the HiTox strain, trafficking of these substrates was completely halted before the protein reached the Golgi, and in the IntTox strain it was reduced (Cooper et al. 2006 and A.A. Cooper, P.K. Auluck, J.M. McCaffery, L.J. Su, and S. Lindquist, unpublished data).

2.3 *High-Throughput Genetic Overexpression Screen*

High-throughput screens in yeast offer approaches for finding druggable targets that are not possible in any other system, including genome-wide combinatorial analysis and the ability to detect gain-, loss-, and change-of-function modifiers. We and others have used and continue to use a variety of different screens to identify factors that influence α-syn toxicity.

Recently, in collaboration with the Harvard Institute of Proteomics (HIP), we completed a genome-wide screen for genes whose overexpression would either enhance or suppress α-syn toxicity in our model (Cooper et al. 2006; Gitler et al., submitted for publication). To identify these toxicity modifiers, we arrayed a library of ∼5000 galactose-regulated, sequence-verified open reading frames (ORFs; the Yeast FLEXGene collection) in the IntTox strain in 96-well plates. Performing this screen in the IntTox strain allowed us to screen simultaneously for genes that would either suppress or enhance the toxicity of α-syn (Fig. 1C). In this screen, both α-syn and the genes from the library were inducibly expressed using the *GAL1* promoter. We were therefore able to perform all manipulations prior to the actual screen in non-inducing conditions, reducing the likelihood that spontaneous, unwanted changes in the genome would accumulate during routine growth and maintenance of the cells. Cells were stamped onto galactose medium to induce α-syn and the ORFs, and growth of the cells was monitored. Suppression of α-syn toxicity resulted in increased growth; enhancement resulted in decreased growth (Fig. 1C).

Given the strong dosage sensitivity of toxicity to α-syn protein levels, we isolated, as expected, genes that simply down-regulate expression from the *GAL1* promoter. Aside from those that served as controls for the screen's efficacy, we identified in this screen over 50 genes that modified α-syn toxicity by enhancing or suppressing it. In keeping with the complexity of all the different factors associated with PD, these genes fell into several different classes, including not only vesicle trafficking but also protein phosphorylation, nitrosative stress, osmolyte synthesis, and metal ion transport (Cooper et al. 2006; Yeger-Lotem et al., submitted for publication).

The largest class of genetic modifiers did comprise genes involved in vesicular trafficking, specifically at the ER-Golgi step (Fig. 1D; Cooper et al. 2006). These hits were logically consistent with what is known about the directionality of vesicle transport: the suppressors were genes predicted to promote anterograde transport (from ER to Golgi) and the enhancers were genes predicted to inhibit it. For example, *YPT1*, a Rab GTPase that promotes the movement of vesicles from ER to Golgi, was one of the first and strongest suppressors isolated. And *GYP8*, a GTPase-activating protein that converts Rab family members from their active GTP-bound state to their inactive GDP-bound state, was a toxicity enhancer. Furthermore, the general functions of these hits suggest that α-syn is likely to be inhibiting the docking or fusion of vesicles to the Golgi rather than inhibiting vesicle budding from the ER (Cooper et al. 2006).

In addition to proteins involved with vesicle trafficking, our genetic screen found three genes for metal ion transporters that modify α-syn toxicity (Cooper et al. 2006;

Gitler et al. submitted for publication). Pmr1p, a strong toxicity enhancer (Cooper et al. 2006), is a secretory pathway ion pump that transports Mn^{2+} and Ca^{2+} ions from the cytoplasm into the Golgi (Durr et al. 1998), and its effects on α-syn toxicity have been confirmed in the *C. elegans* DA neuron system (G.A. Caldwell and S. Lindquist, unpublished results). As an enhancer of toxicity that can be potentially inhibited by small-molecule drug candidates, Pmr1p provides an attractive screening target. An additional yeast metal-ion transporter, Ccc1p, was found to suppress the toxicity of α-syn. This transporter sequesters Fe^{2+} and Mn^{2+} ions into the vacuole of yeast cells (Lapinskas et al. 1996) and suppresses mitochondrial damage in a yeast model of Friedreich's ataxia by limiting mitochondrial iron uptake (Chen and Kaplan 2000). It has been known for some time that patients with PD have higher levels of iron in the substantia nigra (Martin et al. 2008). Mice exposed to iron-chelating drugs or engineered to express natural iron-binding proteins are protected from MPTP toxicity (Kaur et al. 2003). Environmental exposure to manganese has also long been associated with increased risk of developing a Parkinson-like syndrome (Weiss 2006). It is therefore striking that, of the dozens of metal transporters that were in the original library, two of the three hits were associated with the transport of iron and manganese.

The third transporter, YOR291w (Gitler et al., submitted for publication), is a trans-membrane ATPase with unknown specificity. The human homolog of this highly conserved protein, ATP13A2, also known as PARK9, is lysosomal, expressed predominantly in neurons, and believed to couple the hydrolysis of ATP to the transport of cations across cellular membranes (Ramirez et al. 2006). Remarkably, loss-of-function mutants in *ATP13A2* were recently shown to cause hereditary Parkinsonism with dementia (Ramirez et al. 2006), establishing again that genetic findings from screens done in the yeast model can be highly relevant to human disease.

2.4 α-Syn Disrupts Rab-Mediated Vesicle Trafficking Homeostasis

We used a variety of approaches to investigate the specific nature of the ER-to-Golgi trafficking defect that we observed in the yeast model (Gitler et al. 2008). First, in collaboration with Dr. Charles Barlowe (Dartmouth Medical School), we used a cell-free vesicle fusion assay (Barlowe 1997) to test the ability of purified WT and A30P forms of α-syn to inhibit specifically ER-to-Golgi transport. We found that the addition of WT but not A30P α-syn protein (which does not interact with yeast vesicle membranes) to semi-intact membranes inhibited the transport in a dose-dependent manner, and it specifically inhibited the tethering and/or fusion of vesicles to the ER, not the budding of vesicles from the ER.

Ultrastructural analysis of vesicles in the IntTox and HiTox strains (Fig. 3; in collaboration with Dr. JM McCaffery, Johns Hopkins University) established visually that α-syn over-expression in our model caused a dose-dependent block in vesicle trafficking, which in turn resulted in increased vesicle accumulation over

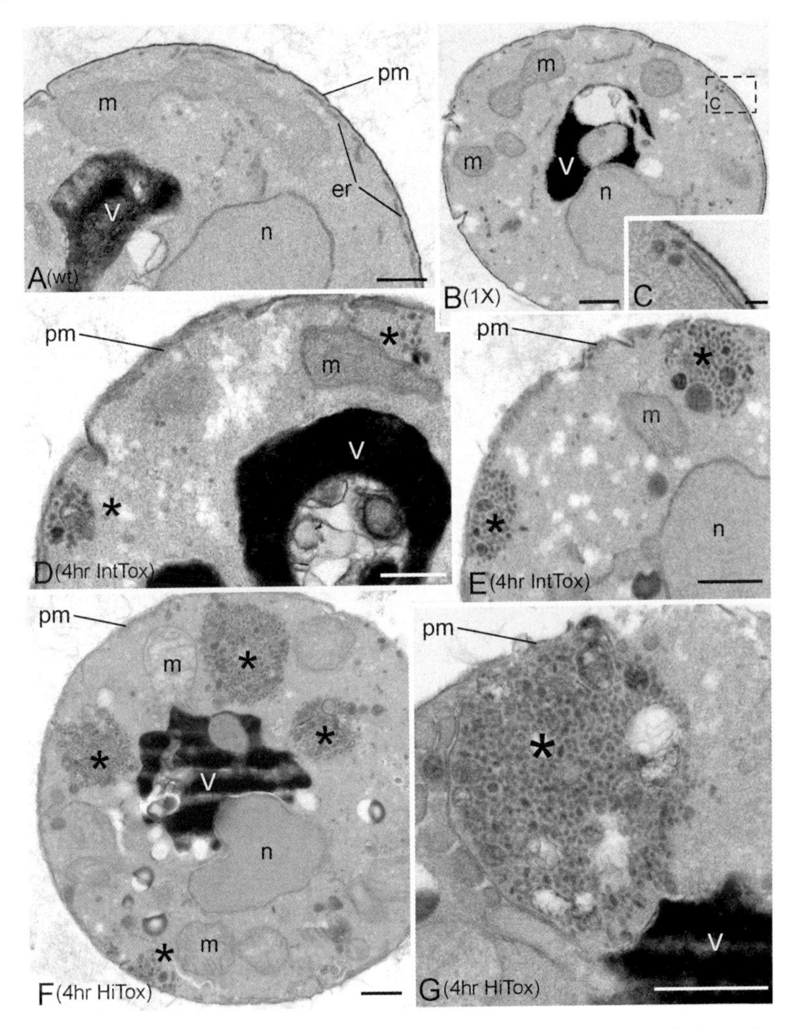

Fig. 3 Electron microscopy reveals a progressive, time-dependent accumulation of α-syn-containing vesicles. (**A**) Vector-only control shows rare, infrequent vesicles similar to those seen with one dose of α-synWT-GFP expression (**B, C**). After four hours of two-dose α-synWT-GFP expression, vesicle accumulation increases to medium/significant (**D, E**) in the IntTox strain or large/massive accumulations (**F, G**) in the HiTox strain. m = mitochondria; n = nucleus; pm = plasma membrane; V = vacuole; er = endoplasmic reticulum; asterisks indicate vesicle clusters in D-G. Scale bars = 0.5 μm. Reprinted from the supplementary online material of Gitler et al. (2008). (Copyright 2008, National Academy of Sciences, U.S.A.)

time. Initially, vesicles accumulated just proximal to the ER and plasma membranes (Fig. 3 B, C), and then, at later time points, they clustered and accumulated in the cell interior in ectopic locations (Fig. 3 D–G). Expressing Ypt1p from the strong *GAL1* promoter rescued the growth defect (Cooper et al. 2006) and strongly reduced α-syn foci in IntTox cells (Gitler et al. 2008). Notably, unlike the IntTox strain, the HiTox stain cannot be rescued by *YPT1*. Furthermore, in this strain these vesicle clusters were clearly heterogeneous and included Golgi markers, indicating that other Rab-mediated trafficking steps had also been disrupted (Fig. 3 F, G and Gitler et al. 2008). α-Syn co-localized with these vesicle clusters and was directly involved in their formation (Gitler et al. 2008).

2.5 Genetic Modifiers of α-Syn Toxicity Validated in Neuronal Models

The pivotal experiment to validate the α-syn yeast model as a discovery platform for the cell biology and pathology of human diseases was to determine whether the genes isolated from yeast screens would have the same effects in whole-animal and neuronal models of the disease. We began by testing the first suppressors in neuronal models (Cooper et al. 2006): the fruit fly *D. melanogaster* (in collaboration with N. Bonini, University of PA), the nematode worm *C. elegans* (in collaboration with G.A. and K.A. Caldwell, University of Alabama), and mixed primary cultures from developing rat midbrain (in collaboration with J.-C. Rochet, Purdue University). Indeed, the human homolog of *YPT1/Rab1* suppressed α-syn toxicity in DA neurons of all three models (Cooper et al. 2006).

While the general secretory pathway is conserved between yeast and higher eukaryotes, yeast lacks the final trafficking steps that are mediated by additional members of the Rab family in neurons and used for the regulated release of synaptic vesicles. For example, *Rab3a* is highly expressed at pre-synaptic sites in neurons (Stettler et al. 1994; Gurkan et al. 2005) and plays an important role in active transport and docking of neurotransmitter vesicles prior to their regulated membrane fusion and release from synaptic terminals (Geppert et al. 1994; Leenders et al. 2001). *Rab8a* is the member of the Rab family with the highest sequence homology to *Rab1* (Gitler et al. 2008) but it also functions in post-Golgi trafficking (Huber et al. 1993). We considered the possibility that dopaminergic neurons might be especially sensitive to disturbances in these later vesicle trafficking steps because of the propensity of unsequestered dopamine to form dangerous ROS. We therefore tested three Rab family members in nematode and rat midbrain culture models (Fig. 4; Gitler et al. 2008) and found that expression of mouse *Rab1* and human *RABs 3A* and *8A* substantially ameliorated α-syn toxicity in DA neurons (Fig. 4; Gitler et al. 2008). These data suggest that α-syn over-expression interferes with multiple steps in neuronal vesicle trafficking and that these individual steps share common genetic mechanisms with each other and with yeast. Thus, data from yeast are consistent with data from mammalian systems that suggest that α-syn acts to reduce the fusion of vesicles with acceptor membranes (Larsen et al. 2006).

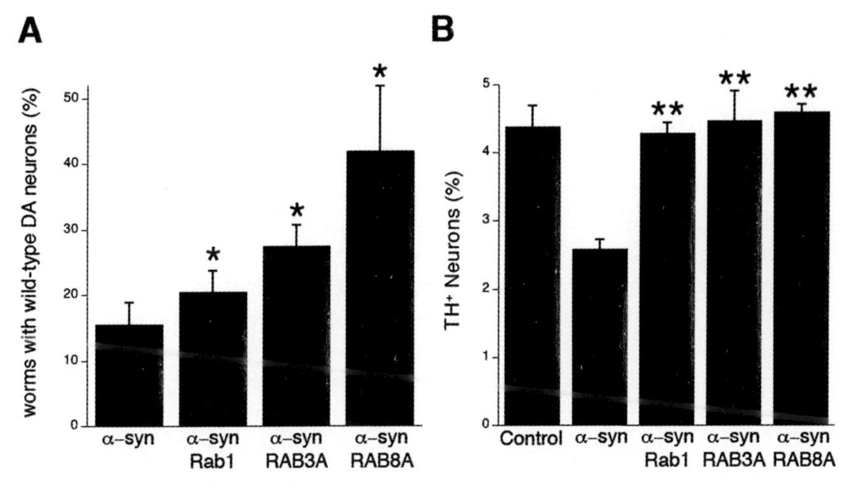

Fig. 4 Rab1, RAB3A, and RAB8A protect against α-syn-induced dopaminergic neuron loss. (**A**) Multiple Rab GTPases ameliorate α-syn-induced neurodegeneration in *C. elegans*. DA neurons of seven-day-old transgenic nematodes overexpressing α-syn along with Rab1, RAB3A, or RAB8A were analyzed. Each Rab tested significantly suppressed α-syn toxicity in worm DA neurons (*, $P < 0.05$, Student's t test). For each gene tested, three transgenic lines were analyzed; a worm was scored as WT when all six anterior DA neurons (four CEP and two ADE neurons) were intact. (**B**) Primary rat midbrain cultures were transduced with A53T lentivirus (multiplicity of infection = 5) in the absence or presence of lentivirus encoding RAB3A, RAB8A, or Rab1 (multiplicity of infection of each Rab virus = 2). Control cells were incubated in the absence of lentivirus. Dopaminergic cell viability was determined by staining with antibodies specific for MAP2 and TH and is expressed as the percentage of MAP2-positive neurons that were also TH-positive (two to three independent experiments; at least 100 cells counted per experiment for each treatment). The data are plotted as the mean +/− SEM. **, $P < 0.001$ vs. A53T virus alone, one-way ANOVA with Newman–Keuls post-test. Reprinted from Gitler et al. (2008). (Copyright 2008, National Academy of Sciences, U.S.A.)

2.6 Chemical Screen Yields Compounds that Rescue Dopaminergic Neurons

We also used our yeast α-syn model to perform a high-throughput chemical screen for small molecules that ameliorate α-syn toxicity (Su et al., submitted for publication). Over 115,000 compounds from various collections, including commercial libraries, natural products, and NCI collections, were screened for their ability to restore growth in yeast cells expressing toxic levels of α-syn. After re-testing, four structurally related compounds were found to restore growth (e.g., Fig. 1E) and antagonize α-syn-mediated inclusion formation and α-syn toxicity at low micromolar concentrations (Su et al., submitted for publicaiton). They also restored α-syn membrane localization and significantly rescued the ER-to-Golgi vesicular trafficking defect, as measured by trafficking of the CPY substrate (Su et al., submitted for publication). As a control for specificity, these lead compounds did not rescue growth in a yeast model expressing polyQ huntingtin exon I, the protein associated with human Huntington's disease.

We tested the most potent compounds from this screen in two established models of α-syn toxicity where genetic rescue had also been demonstrated: DA neurons in α-syn transgenic *C. elegans* and mixed primary cultures of rat midbrain killed either by transduction with human mutant α-syn or addition of the mitochondrial toxin rotenone (G.A. Caldwell, J.-C. Rochet, and S. Lindquist, unpublished; Su et al., submitted for publication). The compounds rescued DA neurons in α-syn transgenic worms, and, unlike *YPT1/Rab1*, they also antagonized toxicity mediated by both these insults in the cultured rat neurons (Su et al., submitted for publication). Rotenone is a mitochondrial poison that interferes with the electron transport chain and can produce toxicities similar to those seen in PD (reviewed in Jenner 2001). Transcriptional profiling (discussed below) and electron microscopy also indicated that α-syn caused mitochondrial damage in the yeast model. Taken together with the connections between mitochondrial stress and PD itself, these results suggest that the mechanism of action for these compounds is central to PD pathobiology. The specificity of the compounds suggests that they may bind to a highly specific cellular target. The target, however, is unlikely to be α-syn itself; the compounds have no effect on the conformational states of purified α-syn (L.J. Su and S. Lindquist, data not shown).

2.7 Transcriptional Profiling of Yeast that Over-Express α-Syn

To identify gene expression changes during the lag period between α-syn induction and cell death, we performed microarray analysis comparing the HiTox strain, the single-dose *URA3* strain, and the empty vector control strain at zero, two, four and six hours after the induction of α-syn. We compared changes in gene expression with the pathobiology undergone by α-syn-expressing yeast cells over these time points. We also examined the effects of the toxicity-rescuing compounds from our chemical screen on the transcription profile. As previously described, after the induction of α-syn in yeast, there is a progressive increase in α-syn expression and concomitant specific alterations in its localization, from its presence at the plasma membrane (at two hours) to the formation of inclusions (at four hours), to a culmination in substantial cell death by six hours (Cooper et al. 2006; Gitler et al. 2008; Su et al., submitted for publication; Yeger-Lotem et al., submitted for publication).

We found that genes involved in mitochondrial processes and ER stress are perturbed by α-syn even at two hours post-induction. The earliest biological defects observed in two-dose α-syn cells were vesicle accumulations and vesicular trafficking defects at three hours post-induction (Cooper et al. 2006; Gitler et al. 2008), consistent with these transcriptional findings. By four hours post-induction, cells expressing two doses of α-syn differentially expressed numerous genes in classes highly enriched for vesicle trafficking, ER stress, sterol biosynthesis, and mitochondrial function. Consistent with the substantial but not complete rescue of cellular toxicity seen with our active compounds, treated HiTox cells showed a much-reduced transcriptional response. With an inactive molecule, the profile was largely similar to that of the untreated HiTox profile (Su et al., submitted for publication; Yeger-Lotem et al., submitted for publication).

Our results in yeast showed a large degree of overlap with the functional categories affected in the published transcriptional profile of a *Drosophila* PD model (Scherzer et al. 2003). These results are also consistent with reports in mammalian PD models and in humans, wherein there has been a long-standing realization that there is a relationship between PD and mitochondrial stress, the formation of ROS, and nitrosative damage (Abou-Sleiman et al. 2006; Uehara et al. 2006). Most importantly, the transcriptional results strongly correlated with what we have learned from the cell biology of our own model system, which shows defective mitochondrial morphology and abnormal mitochondrial DNA, accumulates damaging ROS and undergoes significant oxidative and nitrosative stress, as measured by reactivity to fluorescent probes and antibodies against nitrosylated proteins (Su et al., submitted for publication), before and during the onset of cell death.

2.8 ResponseNet Analysis Bridges Genetic and Transcriptional Data

The two high-throughput methods discussed in detail above, genetic screening and transcriptional profiling, have been extensively used to study cellular responses to perturbations in many different systems. Yet it has been noted that, remarkably, the overlap between these data sets is surprisingly small for the same perturbation (Yeger-Lotem et al., submitted for publication; e.g., compare Begley et al. 2002; Workman et al. 2006). We hypothesized that integrating the two types of data could reveal a more detailed molecular description of disease mechanisms, and we recently developed ResponseNet (Fig. 5), a novel computational approach that uses flow algorithms to bridge the gap between genetic and transcriptional data by known molecular interactions (Yeger-Lotem et al., submitted for publication).

When we applied this approach to data obtained from our α-syn yeast model, the network generated suggested that the heat-shock and target of rapamycin (TOR) pathways would suppress α-syn toxicity. It also indicated that sterol biosynthesis was altered when α-syn was expressed, similar to the result of treatment with the commonly used cholesterol-lowering drugs, statins (Yeger-Lotem et al., submitted for publication). Importantly, ResponseNet and our high-throughput genetic analysis suggest two points in the sterol pathway at which the action of drugs might affect α-syn toxicity, neither of which is cholesterol itself. First, the sterol biosynthesis pathway is required for the production of farensyl groups, which are in turn required for the function of Rabs in vesicle trafficking. Second, the pathway leads to the production of ubiquinone, which is required for complex 1 function in mitochondria. These pathway findings were experimentally validated in the yeast model (Yeger-Lotem et al., submitted for publication) and are supported by results from

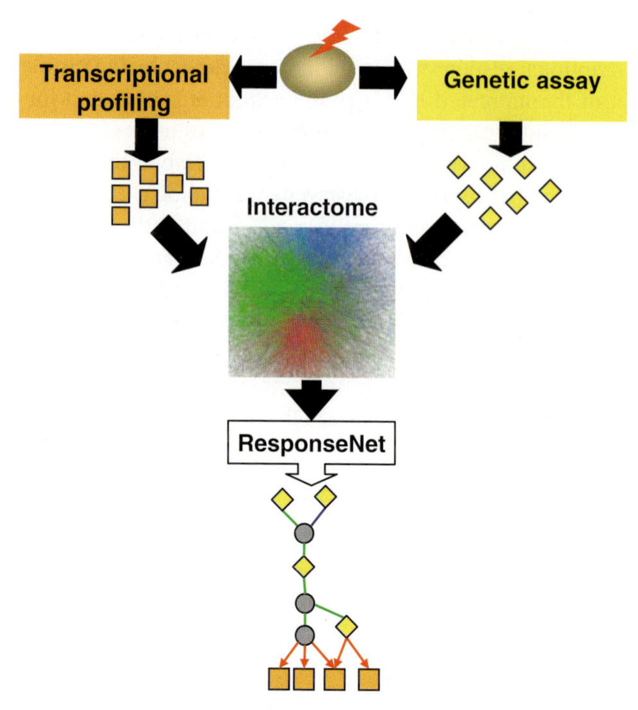

Fig. 5 The ResponseNet algorithm finds paths in the interactome through which a subset of the genetic data may regulate the transcriptional response. Nodes represent proteins and genes, and edges represent their interactions. Diamond shaped nodes represent genetic data (hits from screens), rectangular nodes represent transcriptional data (differentially expressed genes in microarrays), and circular nodes represent intermediate proteins on the paths that link genetic and transcriptional data. The regulation may be direct, as when transcription factors regulating the response are part of the genetic data, or indirect via intermediate proteins

whole-animal α-syn models (e.g., Auluck et al. 2005); most importantly, the sterol pathway results are consistent with recent findings in PD patient populations (Huang et al. 2007). The discovery, using ResponseNet, of the involvement of these pathways substantiates our approach as a powerful tool for interpretation of experimental PD data.

3 Conclusions

We have established simple cellular models for synucleinopathies and other diseases by over-expressing the corresponding human disease-associated proteins in yeast. These yeast cells can be used as a high-throughput discovery platform for factors that modify the toxicity of proteins that are prone to misfolding. Because neurodegenerative diseases often preferentially affect specific neuronal populations, it was, until recently, a commonly held belief that these diseases could only be fruitfully studied in specific neuronal models. However, in at least some cases disease specificity may be the result of a subset of neurons being more vulnerable to general cellular defects that can be profitably studied in yeast. We can take advantage of the extensive genomic and proteomic interaction databases available for yeast to analyze and uncover new roles for known, conserved biochemical pathways that lead to toxicity. Thus, this simple high-throughput approach should be broadly applicable to other cellular processes and other diseases.

When we have conducted screens in multiple models, for example in our models of synucleinopathies and Huntington disease, we have identified disparate sets of hits. Moreover, chemical compounds that ameliorate the toxicity of α-syn do not relieve the toxicity of htt, and vice versa. This lack of overlap establishes that the toxicities observed are not due to non-specific effects of misfolded protein in general but rather to features of the proteins' mis-functioning that interface with cell biology in specific ways. It seems likely that individual disease models will be needed to identify unique therapeutic approaches and meet the growing challenges of an aging population.

Acknowledgements We thank Drs. Esti Yeger-Lotem, Linhui Julie Su, and other members of the Lindquist lab for their thoughtful comments and suggestions. This work was funded in part by Udall program grant NS038372 from the NIH, a Target Validation grant from the Michael J. Fox Foundation for Parkinson's Research, support from the Whitehead Institute Regenerative Biology Initiative to S.L., and support from the HHMI to K.L.A. S.L. is an Investigator of the Howard Hughes Medical Institute and a Fellow of the Radcliffe Institute for Advanced Study at Harvard University.

References

Abou-Sleiman PM, Muqit MM, Wood NW (2006) Expanding insights of mitochondrial dysfunction in Parkinson's disease. Nature Rev Neurosci 7:207–219

Anfinsen CB (1973) Principles that govern the folding of protein chains. Science 181:223–230

Auluck PK, Meulener MC, Bonini NM (2005) Mechanisms of suppression of {alpha}-synuclein neurotoxicity by geldanamycin in Drosophila. J Biol Chem 280:2873–2878

Barlowe C (1997) Coupled ER to Golgi transport reconstituted with purified cytosolic proteins. J Cell Biol 139:1097–1108

Begley TJ, Rosenbach AS, Ideker T, Samson LD (2002) Damage recovery pathways in Saccharomyces cerevisiae revealed by genomic phenotyping and interactome mapping. Mol Cancer Res 1:103–112

Causier B (2004) Studying the interactome with the yeast two-hybrid system and mass spectrometry. Mass Spectrom Rev 23:350–367

Chen OS, Kaplan J (2000) CCC1 suppresses mitochondrial damage in the yeast model of Friedreich's ataxia by limiting mitochondrial iron accumulation. J Biol Chem 275:7626–7632

Cooper AA, Gitler AD, Cashikar A, Haynes CM, Hill KJ, Bhullar B, Liu K, Xu K, Strathearn KE, Liu F, Cao S, Caldwell KA, Caldwell GA, Marsischky G, Kolodner RD, Labaer J, Rochet JC, Bonini NM, Lindquist S (2006) Alpha-synuclein blocks ER-Golgi traffic and Rab1 rescues neuron loss in Parkinson's models. Science 313:324–328

Dauer W, Przedborski S (2003) Parkinson's disease: mechanisms and models. Neuron 39:889–909 den Jager WA (1969) Sphingomyelin in Lewy inclusion bodies in Parkinson's disease. Arch Neurol 21:615

Durr G, Strayle J, Plemper R, Elbs S, Klee SK, Catty P, Wolf DH, Rudolph HK (1998) The medial-Golgi ion pump Pmr1 supplies the yeast secretory pathway with Ca2+ and Mn2+ required for glycosylation, sorting, and endoplasmic reticulum-associated protein degradation. Mol Biol Cell 9:1149–1162

Farrer M, Maraganore DM, Lockhart P, Singleton A, Lesnick TG, de Andrade M, West A, de Silva R, Hardy J, Hernandez D (2001) alpha-Synuclein gene haplotypes are associated with Parkinson's disease. Human Mol Genet 10(17):1847–1851

Gai WP, Yuan HX, Li XQ, Power JT, Blumbergs PC, Jensen PH (2000) In situ and in vitro study of colocalization and segregation of alpha-synuclein, ubiquitin, and lipids in Lewy bodies. Exp Neurol 166:324–333

Gasch AP, Spellman PT, Kao CM, Carmel-Harel O, Eisen MB, Storz G, Botstein D, Brown PO, (2000) Genomic expression programs in the response of yeast cells to environmental changes Mol Biol Cell 11:4241–4257

Geppert M, Bolshakov VY, Siegelbaum SA, Takei K, De Camilli P, Hammer RE, Sudhof TC, (1994) The role of Rab3A in neurotransmitter release. Nature 369:493–497

Giasson BI, Duda JE, Murray IV, Chen Q, Souza JM, Hurtig HI, Ischiropoulos H, Trojanowski JQ, Lee VM (2000) Oxidative damage linked to neurodegeneration by selective a-synuclein nitration in synucleinopathy lesions. Science 290:985–989

Gitler AD, Bevis BJ, Shorter J, Strathearn KE, Hamamichi S, Su LJ, Caldwell KA, Caldwell GA, Rochet JC, McCaffery JM, Barlowe C, Lindquist S (2008) The Parkinson's disease protein alpha-synuclein disrupts cellular Rab homeostasis. Proc Natl Acad Sci USA 105:145–150

Gurkan C, Lapp H, Alory C, Su AI, Hogenesch JB, Balch WE (2005) Large-scale profiling of Rab GTPase trafficking networks: the membrome. Mol Biol Cell 16:3847–3864

Huang X, Chen H, Miller WC, Mailman RB, Woodard JL, Chen PC, Xiang D, Murrow RW, Wang YZ Poole C (2007) Lower low-density lipoprotein cholesterol levels are associated with Parkinson's disease. Mov Disord 22:377–381

Huber LA, Pimplikar S, Parton RG, Virta H, Zerial M, Simons K (1993) Rab8, a small GTPase involved in vesicular traffic between the TGN and the basolateral plasma membrane. J Cell Biol 123:35–45

Irizarry RA, Warren D, Spencer F, Kim IF, Biswal S, Frank BC, Gabrielson E, Garcia JG, Geoghegan J, Germino G, Griffin C, Hilmer SC, Hoffman E, Jedlicka AE, Kawasaki E, Martinez-Murillo F, Morsberger L, Lee H, Petersen D, Quackenbush J, Scott A, Wilson M, Yang Y, Ye SQ, Yu W (2005) Multiple-laboratory comparison of microarray platforms. Nature Methods 2:345–350

Iwatsubo T, Yamaguchi H, Fujimuro M, Yokosawa H, Ihara Y, Trojanowski JQ, Lee VM (1996) Purification and characterization of Lewy bodies from the brains of patients with diffuse Lewy body disease. Am J Pathol 148:1517–1529

Jenner P (2001) Parkinson's disease, pesticides and mitochondrial dysfunction. Trends Neurosci 24(5):245–247

Jo E, McLaurin J, Yip CM, St George-Hyslop P, Fraser PE (2000) alpha-Synuclein membrane interactions and lipid specificity. J Biol Chem 275:34328–34334

Kaur D, Yantiri F, Rajagopalan S, Kumar J, Mo JQ, Boonplueang R, Viswanath V, Jacobs R, Yang L, Beal MF, DiMonte D, Volitaskis I, Ellerby L, Cherny RA, Bush AI, Andersen JK (2003) Genetic or pharmacological iron chelation prevents MPTP-induced neurotoxicity in vivo: a novel therapy for Parkinson's disease. Neuron 37:899–909

Kolkman A, Slijper M, Heck AJ (2005) Development and application of proteomics technologies in Saccharomyces cerevisiae. Trends Biotechnol 23:598–604

Lapinskas PJ, Lin SJ, Culotta VC (1996) The role of the Saccharomyces cerevisiae CCC1 gene in the homeostasis of manganese ions. Mol Microbiol 21:519–528

Larsen KE, Schmitz Y, Troyer MD, Mosharov E, Dietrich P, Quazi AZ, Savalle M, Nemani V, Chaudhry FA, Edwards RH, Stefanis L, Sulzer D (2006) Alpha-synuclein overexpression in

PC12 and chromaffin cells impairs catecholamine release by interfering with a late step in exocytosis. J Neurosci 26:11915–11922

Leenders AG, Lopes da Silva FH, Ghijsen WE, Verhage M (2001) Rab3a is involved in transport of synaptic vesicles to the active zone in mouse brain nerve terminals. Mol Biol Cell 12: 3095–3102

Martin WR, Wieler M, Gee M (2008) Midbrain iron content in early Parkinson disease: A potential biomarker of disease status. Neurology 70:1411–1417.

Outeiro TF, Lindquist S (2003) Yeast cells provide insight into alpha-synuclein biology and pathobiology. Science 302:1772–1775

Ramirez A, Heimbach A, Grundemann J, Stiller B, Hampshire D, Cid LP, Goebel I, Mubaidin AF, Wriekat AL, Roeper J, Al-Din A, Hillmer AM, Karsak M, Liss B, Woods CG, Behrens MI, Kubisch C (2006) Hereditary parkinsonism with dementia is caused by mutations in ATP13A2, encoding a lysosomal type 5 P-type ATPase. Nature Genet 38: 1184–1191

Ramsey CP, Giasson BI (2007) Role of mitochondrial dysfunction in Parkinson's disease: Implications for treatment. Drugs Aging 24:95–105

Scherzer CR, Jensen RV, Gullans SR, Feany MB (2003) Gene expression changes presage neurodegeneration in a Drosophila model of Parkinson's disease. Human MolGenet 12:2457–2466

Sherman F (2002) Getting started with yeast. Methods Enzymol 350: 3–41

Stettler O, Moya KL, Zahraoui A, Tavitian B (1994) Developmental changes in the localization of the synaptic vesicle protein rab3A in rat brain. Neuroscience 62:587–600

Uehara T, Nakamura T, Yao D, Shi ZQ, Gu Z, Ma Y, Masliah E, Nomura Y, Lipton SA (2006) S-nitrosylated protein-disulphide isomerase links protein misfolding to neurodegeneration. Nature 441:513–517

Weiss B (2006) Economic implications of manganese neurotoxicity. Neurotoxicology 27:362–368

Workman CT, Mak HC, McCuine S, Tagne JB, Agarwal M, Ozier O, Begley TJ, Samson LD Ideker T (2006) A systems approach to mapping DNA damage response pathways. Science 312:1054–1059

Tau and Intracellular Transport in Neurons

E.-M. Mandelkow(✉), E. Thies, S. Konzack, and E. Mandelkow

Abstract Among the early changes in the brains of Alzheimer's disease patients is the loss of synapses, which is accompanied by the abnormal phosphorylation of tau protein, its missorting into the somatodendritic compartment of neurons, and its incipient aggregation. The physiological function of tau is to stabilize axonal microtubules, which enables them to carry out their role as tracks for the transport of vesicles and organelles. By implication, perturbations in the functions of tau could be related to the loss of synapses and neuronal degeneration. Cell and transgenic animal models of tauopathy reveal that tau can indeed cause an impairment of transport in neurons. As a result, cell processes of neurons become starved, leading first to the decay of synapses and then to the loss of axons and dendrites.

1 Tau Protein: Properties and Functions

The loss of synapses observed during incipient Alzheimer's disease (AD) corresponds to the beginning loss of memory during the mild cognitive impairment phase (Coleman and Yao 2003). The synapse decay precedes the abnormal protein aggregation of the Aβ peptide in senile plaques or of tau protein in neurofibrillary tangles (Walsh and Selkoe 2004). It has been suspected that the highly elongated structure of neurons is one reason for their vulnerability. Most synapses are distant from the cell body, the major site of protein synthesis, and therefore rely on an efficient transport system. In cells, the traffic system consists of microtubules and microfilaments along which cargoes can be moved (Hollenbeck and Saxton 2005; Baas et al. 2006). This transport is achieved by means of motor proteins that can be subdivided into three classes: myosins (for the microfilament tracks), kinesins and dyneins (for microtubule tracks; Hirokawa and Takemura 2005). The directionality of movement

E.-M. Mandelkow
Max-Planck-Unit for Structural Molecular Biology, 22607 Hamburg, Germany

P. St. George-Hyslop et al. (eds.) *Intracellular Traffic and Neurodegenerative Disorders*,
Research and Perspectives in Alzheimer's Disease,

is determined by the polarity of the tracks and the directionality of the motors. The "plus" ends of microtubules point to the cell periphery, so that plus end-directed motors (kinesin) carry out anterograde transport and minus end-directed motors (dynein) achieve retrograde movements towards the cell body. The "ties" for the tracks are represented by microtubule-associated proteins (MAPs; Cassimeris and Spittle 2001). In neurons, the most important MAPs are MAP2 (mostly dendritic), tau and MAP1b (mostly axonal). The interaction of MAPs with microtubules is controlled by phosphorylation and involves several protein kinases and phosphatases (Mandelkow et al. 2007). Microtubules can assemble from their subunits (α-β-tubulin heterodimers) under the regulation by GTP. Additional control is achieved by MAPs, such as tau, whose detachment can induce microtubule breakdown, and by the microtubule-disassembling proteins, katanin, spastin, or kinesin-13 (Howard and Hyman 2007).

Tau has received attention in the field of several neurodegenerative disorders ("tauopathies") because of its anomalous behavior (Ballatore et al. 2007, Schneider and Mandelkow 2008), which is most conspicuously seen as aggregation into neurofibrillary tangles, consisting of paired helical filaments (PHFs) and straight filaments (Crowther and Goedert 2000; Mandelkow et al. 2007). Tau also becomes highly phosphorylated, missorted into the somatodendritic compartment, partly cleaved by proteases, and otherwise modified (Watanabe et al. 2004; Binder et al. 2005). The H1 haplotype of tau shows a genetic association with certain tauopathies, e.g., progressive supranuclear palsy, corticobasal degeneration, AD and Parkinson disease, which may be caused by a perturbation of tau isoform homeostasis resulting in a relative increase of 4-repeat tau isoforms (inclusion of exon 10) and decrease of N-terminal inserts (especially lack of exon 3; Myers et al. 2007; Caffrey et al. 2007). Biochemically, AD-tau is found to be detached from microtubules and no longer stabilizes microtubules. The consequences are the destabilization of transport tracks and the aggregation of tau in the cytosol, both of which can disrupt intracellular traffic. AD-tau aggregates show a well-defined pattern of spreading in the brain, from the transentorhinal region to the hippocampus and later throughout the cortex. This pattern corresponds to the progression of clinical symptoms from mild cognitive impairment to severe dementia (Braak stages 1–6; Braak and Braak 1991).

The gene of tau (MAPT) is located on chromosome 17; the protein occurs in the CNS as six main isoforms arising from alternative splicing (352–441 residues; Andreadis 2005). The repeat domain (3 or 4 pseudo-repeats of ~31 residues, depending on splice isoforms) and the domains flanking the repeats are responsible for microtubule binding. The repeat domain also forms the core of Alzheimer PHFs (Wille et al. 1992; Novak et al. 1993). The overall character of tau is basic and hydrophilic, due to the many lysine or arginine and polar residues, which makes tau highly soluble, up to the point that tau is heat and acid stable without losing its biological function (Lee et al. 1988). A further consequence is that tau is not compactly folded as most proteins but rather is a natively unfolded protein (Schweers et al. 1994). Several mutations in the tau gene can cause different forms of neurodegeneration (FTDP-17; Ballatore et al. 2007), presumably due to a change in protein function or an altered distribution of isoforms caused by modifications in the pattern of alternative splicing (D'Souza and Schellenberg 2005).

Tau from AD brains is extensively phosphorylated, ~4-fold higher than in normal brain and at numerous sites (Khatoon et al. 1992; Morishima-Kawashima et al. 1995). The consequences are heterogeneous. Phosphorylation at certain sites can affect microtubule binding and/or PHF aggregation; other sites appear to be functionally neutral (Schneider et al. 1999). Phosphorylation at the KXGS motifs in the repeat domain by the kinase MARK strongly disrupts tau-microtubule binding and leads to dynamic microtubules (Drewes et al. 1997). The interplay between tau and MARK becomes particularly noticeable in the case of neurite outgrowth, where activation of MARK has a similar effect as nerve growth factor signalling (Biernat et al. 2002).

The most unusual property of tau in AD is its aggregation, which is counterintuitive because of the high solubility of tau. The aggregation is based on certain hexapeptide motifs in the sequence that have an increased propensity for β-sheet interactions (^{275}VQIINK^{280} and ^{306}VQIVYK^{311}; von Bergen et al. 2000). Therefore, the aggregation of tau is based on an "amyloid" principle, although the major part of the protein remains disordered, even when it is assembled into PHFs. This finding is borne out by recent structural results. X-Ray crystallography reveals that amyloidogenic peptides derived from different proteins form hairpin-like "amyloid spines" that assemble into cross-β-sheets, stabilized by internal hydrophobic interactions and hydrogen bonds and paired by hydrophilic interactions (Sawaya et al. 2007). Nuclear magnetic resonance studies reveal that the amyloidogenic subdomains have an enhanced tendency for extended conformation with β-propensity even in solution, which is stabilized in hairpin-like conformations during fiber assembly (Mukrasch et al. 2005; Andronesi et al. 2008).

2 Tau and Transport Inhibition in Neurons

The traffic systems in neurons can be regulated at different levels, for example at the level of tracks (microtubules, tau), motors (kinesin, dynein), or cargo adaptors (kinesin or dynein light chains or associated proteins), or by posttranslational modifications (phosphorylation; Stokin and Goldstein 2006). In this context, proteins closely related to AD include tau and protein kinases that can regulate tracks, motors, or adaptors. In cells, one observes that elevation of tau causes a stabilization of microtubules as well as a general inhibition of intracellular traffic, particularly in the anterograde direction (Stamer et al. 2002; Fig. 1), which can be explained by the fact that tau inhibits both forward motors (kinesin) and reverse motors (dynein), but the inhibition of kinesin is more pronounced, resulting in a net retrograde bias in the transport of vesicles and organelles (Seitz et al. 2002; Dixit et al. 2008). The observations are consistent with the view that the attachment of motors is obstructed by tau bound to the microtubule tracks. In addition, tau may interact directly with kinesin or dynein motors (Magnani et al. 2007; Cuchillo-Ibanez et al. 2008). The

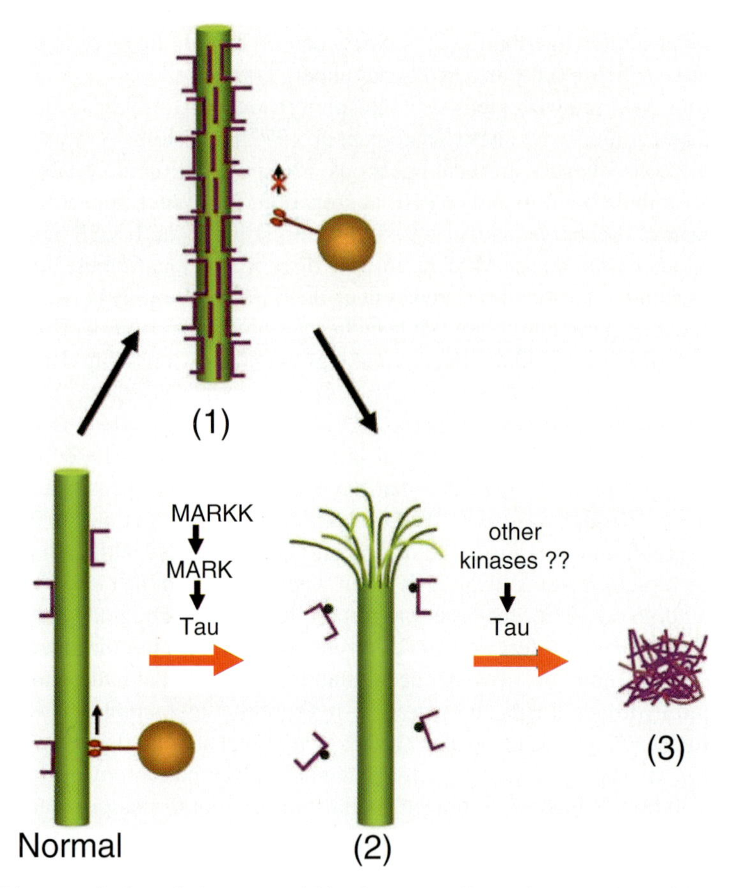

Fig. 1 Diagram of microtubules, tau, and kinesin motors, illustrating several possible modes of dysregulation of the neuronal transport system. Lower left, normal state with intact microtubule tracks, sparse decoration by tau (which suffices for stabilization), and kinesin motor with vesicle cargo moving in the anterograde direction. (**1**) Too much tau bound to the microtubule surface can restrict the access of motor proteins, retard axonal transport, and overstabilize microtubules (leading to insufficient dynamic instability and excess microtubule polymerization). (**2**) Tau may become hyperphosphorylated (e.g., in the repeat domain by the kinase MARK), which causes detachment from microtubules. This may lead to destabilization of microtubules and thus to loss of transport tracks. (**3**) Tau detached from microtubules may aggregate into paired helical filaments, which coalesce into neurofibrillary tangles that obstruct the cell interior

traffic inhibition can be rescued by phosphorylating tau such that it detaches from microtubules, e.g., by the kinase MARK, which phosphorylates the repeat domain (Mandelkow et al. 2004; Thies and Mandelkow 2007; Fig. 2a, b). By the same token, variants of tau that adhere strongly to tau can inhibit traffic more strongly than variants that adhere weakly (Konzack et al. 2007). These features indicate that even elevated normal tau bound to microtubules can perturb the cell's physiological functions, which can become a serious problem for elongated cells such as neurons that are dependent on an efficient transport system. For example, the depletion of

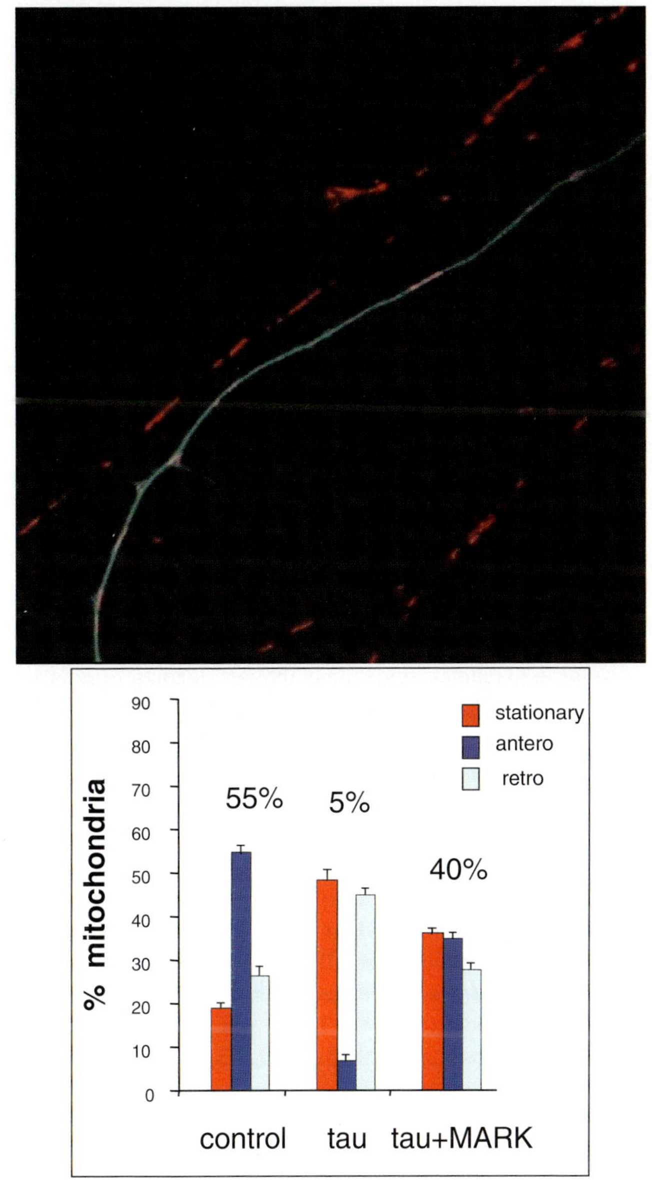

Fig. 2 Transport inhibition by tau in retinal ganglion cell axons. (**a**) Retinal ganglion cell axons growing out from an explant. Mitochondria are stained with MitoTracker Red; one cell is transfected with CFP-tau by adenovirus (blue). There are numerous highly mobile mitochondria in most cells, but the tau-transfected cell has already lost most of its mitochondria, and the remaining ones are almost immobile and are in the process of degenerating. (**b**) Quantification of mitochondrial movements. In control growing axons, the majority of mitochondria move anterogradely (55%). In tau-transfected cells, this fraction drops to 5%, and most mitochondria either move retrogradely or are stationary within the window of observation. When cells are additionally transfected with MARK, the microtubule-bound tau becomes phosphorylated and detaches from microtubules, and anterograde movement of mitochondria is largely restored

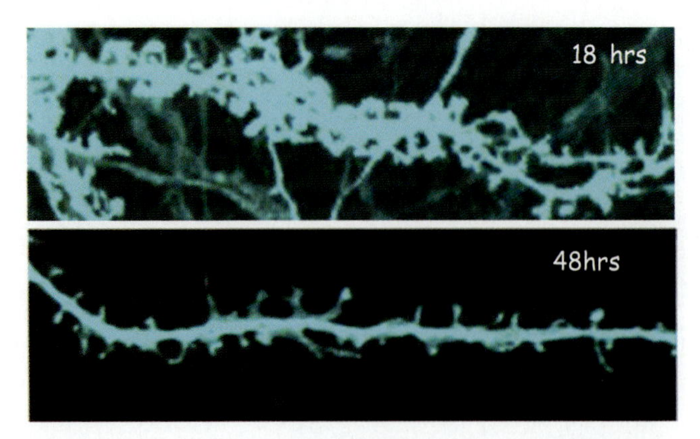

Fig. 3 Decay of dendritic spines and synapses under the influence of tau. Top: Hippocampal neurons were cultured for 25 days in vitro, leading to numerous synapses. The neurons were transfected with CFP-tau (blue), which enters the dendritic compartment, including spines, due to missorting of tau. Bottom: 20 hours later most spines have withered away, concomitant with a loss of energy (ATP), loss or displacement of synaptic markers, and translocation of F-actin into the dendritic shaft (not shown)

organelles such as mitochondria from cell processes by inhibition of anterograde transport will cause deficiencies in local metabolism, leading to reduced adenosine triphosphate (ATP) levels, Ca^{++} buffering capacity, and defense against oxidative stress. These reductions will impair the overall viability of the cells, and among the first victims are the dendritic spines (Thies and Mandelkow 2007), reminiscent of the early decay of synapses in AD (Fig. 3).

A special aspect of tau-induced traffic inhibition is that vesicles carrying amyloid precursor protein (APP) are affected as well, suggesting a potential link between the two major pathological hallmarks in AD. However, there appears to be no direct link to the generation of the Aβ amyloid peptide, and vesicles carrying APP are distinct from those carrying the protease BACE1 (responsible for the first cleavage of APP leading to Aβ), making a direct interaction between APP and BACE1 during transport unlikely (Goldsbury et al. 2006; Lazarov et al. 2005).

While the inhibition of axonal traffic by tau is easily observable in vitro and in cell models, the results from studies on organisms are heterogeneous. Mouse models, where the tau gene is expressed under the pan-neuronal promotor Thy-1, suffer from motor neuron disease, which makes it difficult to test any effects on memory. This disease has been ascribed to the expression of tau in motor neurons that are particularly long and therefore vulnerable to traffic inhibition by tau (Terwel et al. 2002; Götz et al. 2006; Lee et al. 2005). This problem can be circumvented by using promoters restricted to the forebrain, e.g., the CaMKII promoter. Similarly, expression of tau in the axons of flies causes traffic deficits and damage to the neuromuscular junction (Chee et al. 2005; Fulga et al. 2007). In Aplysia, overstabilization of microtubules by tau can take the form of misoriented microtubules that obstruct axons (Shemesh et al. 2008). On the other hand, a tau-induced inhibition of traffic has not been observed in extruded squid giant axoplasm (Morfini et al. 2007) or in vivo in retinal ganglion cells from mice overexpressing tau (Yuan et al. 2008). These variable results may be related to differences in regulatory systems operating in the experimental systems used.

One of the puzzling features of the behavior of tau in neurons is that, on the one hand, it inhibits anterograde traffic of microtubule-dependent cargo (vesicles, organelles, neurofilaments) but, on the other hand, tau itself, when elevated, enters axons and dendrites with apparent ease. Thus, while vesicles and organelles tend to disappear from cell processes, tau moves out, seemingly "against the tide" (Konzack et al. 2007). The solution to this paradox resides in two features of tau. First, anterograde transport of tau occurs on small microtubule fragments that can be transported not only on microtubules but also on actin filaments (Wang and Brown 2002; Baas et al. 2006). In the latter case, a dynein-based movement of tau-tubulin complexes relative to a stationary actin network would achieve the required anterograde directionality (note that dynein is much less affected by tau than kinesin; see above). This finding would explain the observed rates of tau within the slow component b (Scb) of axonal transport (Mercken et al. 1995; Roy et al. 2008). Secondly, tau is much more mobile than anticipated for a "microtubule-associated" protein. It spends only ∼70% of the time on the microtubule, associates and dissociates rapidly (residence time ∼4 sec) and, in the unbound state diffuses freely in the cytoplasm (Konzack et al. 2007). Thus, over periods of ∼days and distances of ∼mm, diffusion appears to be adequate to supply the axon with the required level of tau.

3 Conclusions

In this review we have considered some mechanism by which tau could gain toxic functions, based on the known functions of tau in neurons. They can be summarized as follows:

1. Elevated tau, when bound to the microtubule surface, can inhibit the attachment of motor proteins and thus slow down transport rates in the cell processes of neurons (Stamer et al. 2002; Figs. 1, 2).
2. Elevated tau can overstabilize microtubules and generate excess microtubules in axons and dendrites, with two consequences: the excess microtubules can fill out the cytosolic space and thus prevent the transit of vesicles and organelles (Thies and Mandelkow 2007; Fig. 4).
3. In addition, the overstabilization of microtubules by excess tau can suppress the dynamic instability of microtubules, which is necessary for maintaining the capacity for remodelling the neuronal cytoskeleton (Baas et al. 2006). In this context, it is notable that the 4-repeat tau isoforms (which bind and stabilize microtubules better than the fetal 3-repeat isoforms) suppress microtubule dynamics in a fashion reminiscent of taxol, a microtubule poison used in cancer chemotherapy (Panda et al. 2003).

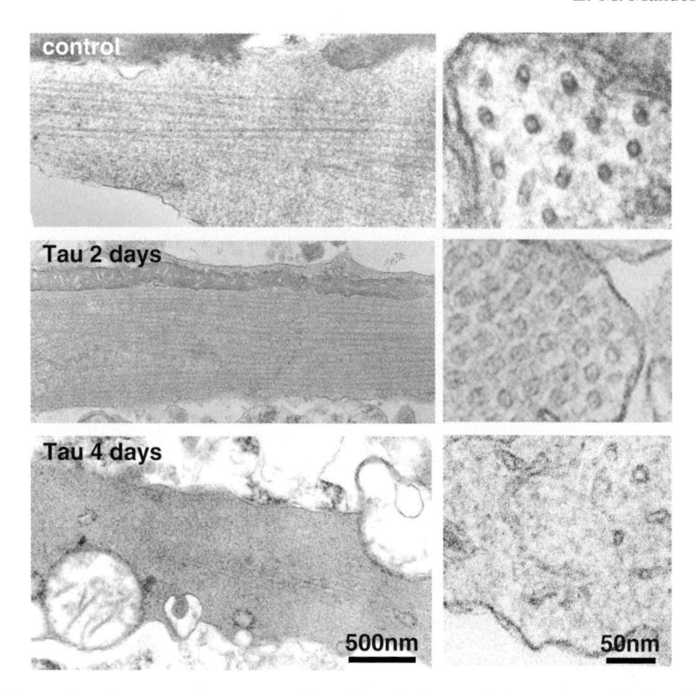

Fig. 4 Thin-section electron microscopy of dendrites before and after tau transfection. In the control cells (top), the microtubules are spaced wide apart and allow passage of transport vesicles and organelles. After two days of transfection with tau, microtubules become more numerous (fourfold) because tubulin synthesis is initially upregulated. Microtubules become more densely spaced, thus blocking transit (note the mitochondrion pushed to the upper side of the cell). After four days, microtubules have mostly disappeared (due to lack of ATP and GTP) and mitochondria are swollen and in a state of degeneration. The tau-induced changes can be halted, at least temporarily, by phosphorylating tau (by kinase MARK) and thereby detaching it from microtubules (Thies and Mandelkow 2007)

4. Tau could interfere with transport by inactivating complexes of motor proteins directly (e.g., kinesin, dynactin; Magnani et al. 2007; Cuchillo-Ibanez et al. 2008).
5. Hyperphosphorylation of tau and subsequent detachment from microtubules could destabilize microtubules, leading to a loss of transport tracks (Zhang et al. 2004; Fig. 1).
6. Tau could interfere with the functions of other cellular proteins, e.g.. the actin network (Fulga et al. 2007), and signalling molecules (e.g.. Pin-1, Lippens et al. 2007; kinases and phosphatases, Stoothoff and Johnson 2005; chaperones, Shimura et al. 2004).

On the level of the tau gene and the mutations known from FTDP-17 and other tauopathies, it is notable that the shift in the splicing pattern generates an imbalance between 4-repeat and 3-repeat tau isoforms in favor of 4-repeat forms. The 4-repeat forms bind and stabilize microtubules more strongly, but aggregate into PHFs less readily than 3-repeat isoforms. This finding is suggestive of a mechanism of toxicity based on blocking traffic and/or suppressing microtubule dynamics (points 1–3 above). Consistent with this, the H1 haplotype of MAPT causes a higher level of 4-repeat tau protein in neurons, which might explain why this haplotype represents a risk factor for several tauopathies (Myers et al. 2007).

In this review we focussed on the question of how tau might disturb the neuronal transport system and thus contribute to neurodegeneration. We have not considered the causes and effects of the abnormal tau aggregation that is prominent in tauopathies. However, it is notable that certain cell and animal models display a strong tau-dependent toxicity that is specifically related to aggregation (Wang et al. 2007; Mocanu et al. 2008). This mode appears to be independent of the mode of transport-related toxicity, consistent with the observation that transport defects and aggregation defects occur at different stages of the AD process (Götz et al. 2006).

Acknowledgement

Work from our laboratory discussed here was supported by the Max-Planck-Gesellschaft (MPG) and the Deutsche Forschungsgemeinschaft (DFG).

References

Andreadis, A (2005) Tau gene alternative splicing: expression patterns, regulation and modulation of function in normal brain and neurodegenerative diseases. Biochim Biophys Acta 1739: 91–103.

Andronesi O, von Bergen M, Biernat J, Seidel K, Heise H, Griesinger C, Mandelkow E, Baldus M (2008) Characterization of Alzheimer's-like paired helical filaments from the core domain of tau protein using solid-state NMR spectroscopy. J Am Chem Soc 130: 5922–5928.

Baas PW, Vidya Nadar C, Myers KA (2006) Axonal transport of microtubules: the long and short of it. Traffic 7:490–498.

Ballatore C, Lee VM, Trojanowski JQ (2007) Tau-mediated neurodegeneration in Alzheimer's disease and related disorders. Nat Rev Neurosci 8:663–672.

Biernat J, Wu YZ, Timm T, Zheng-Fischhöfer Q, Mandelkow E, Meijer L, Mandelkow EM (2002) Protein kinase MARK/PAR-1 is required for neurite outgrowth and establishment of neuronal polarity. Mol Biol Cell 13:4013–4028.

Binder LI, Guillozet-Bongaarts AL, Garcia-Sierra F, Berry RW (2005) Tau, tangles, and Alzheimer's disease. Biochim Biophys Acta 1739:216–223.

Braak H, Braak E (1991) Neuropathological stageing of Alzheimer-related changes. Acta Neuropathol 82:239–259.

Caffrey TM, Joachim C, Wade-Martins R (2007) Haplotype-specific expression of the N-terminal exons 2 and 3 at the human MAPT locus. Neurobiol Aging. 2007 Jun 27. [Epub ahead of print].

Cassimeris L, Spittle C (2001) Regulation of microtubule-associated proteins. Int Rev Cytol 210:163–226.

Chee FC, Mudher A, Cuttle MF, Newman TA, MacKay D, Lovestone S, Shepherd D (2005) Over-expression of tau results in defective synaptic transmission in Drosophila neuromuscular junctions. Neurobiol Dis 20:918–928.

Chen J, Kanai Y, Cowan NJ, Hirokawa N (1992) Projection domains of MAP2 and tau determine spacings between microtubules in dendrites and axons. Nature 360:674–677.

Coleman PD, Yao PJ (2003) Synaptic slaughter in Alzheimer's disease. Neurobiol Aging 24:1023–1027.

Crowther RA, Goedert M (2000) Abnormal tau-containing filaments in neurodegenerative diseases. J Struct Biol 130:271–279.

Cuchillo-Ibanez I, Seereeram A, Byers HL, Leung KY, Ward MA, Anderton BH, Hanger DP (2008). Phosphorylation of tau regulates its axonal transport by controlling its binding to kinesin. FASEB J. Jun 3. [Epub ahead of print].

Dixit R, Ross JL, Goldman YE, Holzbaur EL (2008) Differential regulation of dynein and kinesin motor proteins by tau. Science 319:1086–1089.

D'Souza I, Schellenberg GD (2005) Regulation of tau isoform expression and dementia. Biochim Biophys Acta 1739:104–115.

Drewes G, Ebneth A, Preuss U, Mandelkow EM, Mandelkow E (1997) MARK, a novel family of protein kinases that phosphorylate microtubule-associated proteins and trigger microtubule disruption. Cell 89:297–308.

Fulga TA, Elson-Schwab I, Khurana V, Steinhilb ML, Spires TL, Hyman BT, Feany MB (2007) Abnormal bundling and accumulation of F-actin mediates tau-induced neuronal degeneration in vivo. Nature Cell Biol. 9:139–148.

Goldsbury, C., Mocanu, M., Thies, E., Kaether, C., Haass, C., Keller, P., Biernat, J., Mandelkow, E.-M., Mandelkow, E. (2006) Inhibition of APP trafficking by tau protein does not increase the generation of amyloid-beta. Traffic 7:873–888.

Götz J, Ittner LM, Kins S (2006) Do axonal defects in tau and amyloid precursor protein transgenic animals model axonopathy in Alzheimer's disease? J Neurochem 98:993–1006.

Hirokawa N, Takemura R (2005) Molecular motors and mechanisms of directional transport in neurons. Nature Rev Neurosci 6:201–214.

Hirokawa N, Funakoshi T, Sato-Harada R, Kanai Y (1996) Selective stabilization of tau in axons and microtubule-associated protein 2C in cell bodies and dendrites contributes to polarized localization of cytoskeletal proteins in mature neurons. J Cell Biol 132:667–679.

Hollenbeck PJ, Saxton W (2005) The axonal transport of mitochondria. J Cell Sci 118:5411–5419.

Howard J, Hyman AA (2007) Microtubule polymerases and depolymerases. Curr Opin Cell Biol 19:31–35.

Ishihara T, Hong M, Zhang B, Nakagawa Y, Lee MK, Trojanowski JQ, Lee VM (1999) Age-dependent emergence and progression of a tauopathy in transgenic mice overexpressing the shortest human tau isoform. Neuron 24:751–762.

Kanai Y, Hirokawa N (1995) Sorting mechanisms of tau and MAP2 in neurons: suppressed axonal transit of MAP2 and locally regulated microtubule binding. Neuron. 14:421–432.

Khatoon S, Grundke-Iqbal I, Iqbal K (1992) Brain levels of microtubule-associated protein tau are elevated in Alzheimer's disease: a radioimmuno-slot-blot assay for nanograms of the protein. J Neurochem 59:750–753.

Konzack S, Thies E, Marx A, Mandelkow E-M, Mandelkow E (2007) Swimming against the tide: mobility of the microtubule-associated protein tau in neurons. J. Neurosci 27:9916–9927.

Lazarov O, Morfini GA, Lee EB, Farah MH, Szodorai A, DeBoer S, Koliatsos VE, Kins S, Lee VM, Wong PC, Price DL, Brady ST, Sisodia SS (2005) Axonal transport, amyloid precursor protein, kinesin-1, and the processing apparatus: revisited. J Neurosci 25:2386–2395.

Lee G, Cowan N, Kirschner M (1988) The primary structure and heterogeneity of tau protein from mouse brain. Science 239:285–288.

Lee VM, Kenyon TK, Trojanowski JQ (2005) Transgenic animal models of tauopathies. Biochim Biophys Acta 1739:251–259.

Lippens G, Landrieu I, Smet C (2007) Molecular mechanisms of the phospho-dependent prolyl cis/trans isomerase Pin1. FEBS J. 274:5211–5222.

Magnani E, Fan J, Gasparini L, Golding M, Williams M, Schiavo G, Goedert M, Amos LA, Spillantini MG (2007) Interaction of tau protein with the dynactin complex. EMBO J 26:4546–4554.

Mandelkow, E., von Bergen, M., Biernat, J., Mandelkow, E.-M. (2007). Structural principles of tau and Alzheimer paired helical filaments. Brain Pathol 17:83–90.
Mandelkow E-M, Thies E, Trinczek B, Biernat B, Mandelkow E (2004) MARK/PAR1 kinase is a regulator of microtubule-dependent transport in axons. J Cell Biol 167:99–110.
Mercken M, Fischer I, Kosik KS, Nixon RA (1995) Three distinct axonal transport rates for tau, tubulin, and other microtubule-associated proteins: evidence for dynamic interactions of tau with microtubules in vivo. J Neurosci 15:8259–8267.
Mocanu M, Nissen A, Eckermann K, Khlistunova I, Biernat J, Drexler D, Petrova O, Schönig K, Bujard H, Mandelkow E, Zhou L, Rune G, Mandelkow EM (2008) The potential for beta structure in the repeat domain of Tau protein determines aggregation, synaptic decay, neuronal loss, and co-assembly with endogenous Tau in inducible mouse models of tauopathy. J Neurosci 28:737–748.
Morfini G, Pigino G, Mizuno N, Kikkawa M, Brady ST (2007) Tau binding to microtubules does not directly affect microtubule-based vesicle motility. J Neurosci Res 85:2620–2630.
Morishima-Kawashima M, Hasegawa M, Takio K, Suzuki M, Yoshida H, Titani K, Ihara Y (1995). Proline-directed and non-proline-directed phosphorylation of PHF-tau. J Biol Chem 270:823–829.
Mukrasch MD, Biernat J, von Bergen M, Griesinger C, Mandelkow E, Zweckstetter M (2005) Sites of tau important for aggregation populate beta structure and bind to microtubules and polyanions. J Biol Chem 280:24978–24986.
Myers AJ, Pittman AM, Zhao AS, Rohrer K, Kaleem M, Marlowe L, Lees A, Leung D, McKeith IG, Perry RH, Morris CM, Trojanowski JQ, Clark C, Karlawish J, Arnold S, Forman MS, Van Deerlin V, de Silva R, Hardy J (2007) The MAPT H1c risk haplotype is associated with increased expression of tau and especially of 4 repeat containing transcripts. Neurobiol Dis 25:561–570.
Novak M, Kabat J, Wischik CM (1993) Molecular characterization of the minimal protease resistant tau unit of the Alzheimer's disease paired helical filament. EMBO J 12:365–370.
Panda D, Samuel JC, Massie M, Feinstein SC, Wilson L (2003) Differential regulation of microtubule dynamics by three- and four-repeat tau: implications for the onset of neurodegenerative disease. Proc Natl Acad Sci USA 100:9548–9553.
Roy S, Winton MJ, Black MM, Trojanowski JQ, Lee VM (2008) Cytoskeletal requirements in axonal transport of slow component-b. J Neurosci 28:5248–5256.
Sawaya MR, Sambashivan S, Nelson R, Ivanova MI, Sievers SA, Apostol MI, Thompson MJ, Balbirnie M, Wiltzius JJ, McFarlane HT, Madsen A, Riekel C, Eisenberg D (2007) Atomic structures of amyloid cross-beta spines reveal varied steric zippers. Nature 447:453–457.
Schneider A, Mandelkow E (2008) Tau-based treatment strategies in neurodegenerative diseases. Neurotherapeutics 5:443–457.
Schneider A, Biernat J, von Bergen M, Mandelkow E, Mandelkow E-M (1999) Phosphorylation that detaches tau protein from microtubules (Ser262, Ser214) also protects it against aggregation into Alzheimer paired helical filaments. Biochemistry 38:3549–3558.
Schweers O, Schonbrunn-Hanebeck E, Marx A, Mandelkow E (1994) Structural studies of tau protein and Alzheimer paired helical filaments show no evidence for beta-structure. J Biol Chem 269:24290–24297.
Seitz A, Kojima H, Oiwa K, Mandelkow E-M, Song Y-H, Mandelkow E (2002) Single-molecule investigation of the interference between kinesin and tau on microtubules. EMBO J 21:4896–4905.
Shemesh OA, Erez H, Ginzburg I, Spira ME (2008) Tau-induced traffic jams reflect organelles accumulation at points of microtubule polar mismatching. Traffic 9:458–471.
Shimura H, Schwartz D, Gygi SP, Kosik KS (2004) CHIP-Hsc70 complex ubiquitinates phosphorylated tau and enhances cell survival. J Biol Chem 279:4869–4876.
Stamer K, Vogel R, Thies E, Mandelkow E, Mandelkow E-M (2002) Tau blocks traffic of organelles, neurofilaments, and APP-vesicles in neurons and enhances oxidative stress. J Cell Biol 156:1051–1063.

Stokin GB, Goldstein LS (2006) Axonal transport and Alzheimer's disease. Annu Rev Biochem. 75:607–627.
Stoothoff WH, Johnson GV (2005) Tau phosphorylation: physiological and pathological consequences. Biochim Biophys Acta 1739:280–297.
Terwel D, Dewachter I, Van Leuven F (2002) Axonal transport, tau protein, and neurodegeneration in Alzheimer's disease. Neuromolecular Med 2:151–165.
Thies E, Mandelkow E-M (2007) Missorting of tau in neurons causes degeneration of synapses that can be rescued by MARK2/Par-1. J Neurosci 27:2896–2907.
von Bergen M, Friedhoff P, Biernat J, Heberle J, Mandelkow EM, Mandelkow E (2000) Assembly of tau protein into Alzheimer paired helical filaments depends on a local sequence motif 306-VQIVYK-311 forming beta structure. Proc Natl Acad Sci USA 97:5129–5134.
Walsh DM, Selkoe DJ (2004) Deciphering the molecular basis of memory failure in Alzheimer's disease. Neuron 44:181–193.
Wang L, Brown, A (2002) Rapid movement of microtubules in axons. Curr Biol 12:1496–1501.
Wang YP, Biernat J, Pickhardt M, Mandelkow E, Mandelkow E-M (2007) Stepwise proteolysis liberates tau fragments that nucleate the Alzheimer-like aggregation of full-length tau in a neuronal cell model. Proc Natl Acad Sci USA 104:10252–10257.
Watanabe A, Hong WK, Dohmae N, Takio K, Morishima-Kawashima M, Ihara Y (2004) Molecular aging of tau: disulfide-independent aggregation and non-enzymatic degradation in vitro and in vivo. J Neurochem 90:1302–1311.
Wille H, Drewes G, Biernat J, Mandelkow EM, Mandelkow E (1992) Alzheimer-like paired helical filaments and antiparallel dimers formed from microtubule-associated protein tau in vitro. J Cell Biol.118:573–584.
Yuan A, Kumar A, Peterhoff C, Duff K, Nixon RA (2008) Axonal transport rates in vivo are unaffected by tau deletion or overexpression in mice. J Neurosci 28:1682–1687.
Zhang B, Higuchi M, Yoshiyama Y, Ishihara T, Forman MS, Martinez D, Joyce S, Trojanowski JQ, Lee VM (2004) Retarded axonal transport of R406W mutant tau in transgenic mice with a neurodegenerative tauopathy. J Neurosci 24:4657–4667.

Signaling Between Synapse and Nucleus During Synaptic Plasticity

Kwok-On Lai, Dan Wang, and Kelsey C. Martin(✉)

Abstract The requirement for transcription during synapse formation and long-lasting synaptic plasticity raises two cell biological questions regarding communication between the synapse and nucleus in neurons. First, how are signals transmitted from stimulated synapses to the nucleus to initiate changes in gene expression? Second, how do the products of gene expression function to alter the structure and function of some but not all synapses made by a given neuron? We address these questions in two model systems of synapse formation and synaptic plasticity: cultured sensory-motor neurons from the marine mollusk, *Aplysia californica*, and cultured rodent hippocampal neurons. In studying signaling from synapse to nucleus, we have discovered a role for the importin family of nuclear transporters in carrying signals from synapses to the nucleus during long-term plasticity. Importins are present at synapses, travel to the nucleus following stimuli that elicit transcription, and are required for the long-term plasticity of *Aplysia* sensory-motor synapses. In studying how the products of transcription are targeted to specific synapses within a neuron, we have focused on the role of mRNA localization and regulated translation at the synapse. We have identified hundreds of mRNAs that are present in distal processes of *Aplysia* neurons and in dendrites of rodent hippocampal neurons. We find that localized mRNAs are translated at sites of synaptic contact during synapse formation and following stimulation that elicits long-lasting synaptic plasticity. It will be of interest to determine how these long-range, synapse-to-nucleus signal transduction pathways are altered with aging and during neurodegeneration.

K.C. Martin
Dept. of Psychiatry and Biobehavioral Sciences, Brain Research Institute, Dept of Biological Chemistry, Semel Institute for Neuroscience and Human Behavior, UCLA, BSRB 390B, 615 Charles E. Young Dr. S., Los Angeles, CA 90095-1737

P. St. George-Hyslop et al. (eds.) *Intracellular Traffic and Neurodegenerative Disorders*, Research and Perspectives in Alzheimer's Disease,

1 Introduction

Synaptic plasticity, the process whereby neurons change the structure and function of their connections with experience, provides a mechanism for memory storage in the brain (for recent reviews, see Neves et al. 2008; Bruel-Jungerman et al. 2007; Kim and Linden 2007). In its most general form, the synaptic plasticity hypothesis postulates that memories are stored as increases in the strength and/or number of synaptic contacts between neurons within a given circuit. Supporting this hypothesis, studies in a number of systems have indicated that synaptic plasticity is indeed critical to cognition and memory formation (reviewed in Neves et al. 2008).

Like memory, synaptic plasticity can be divided into short-term and long-term forms, which differ in their requirement for new gene expression (Alberini 1999; Kandel, 2001). Thus, short-term forms of plasticity depend on covalent modifications of existing proteins, whereas long-term plasticity requires new RNA and protein synthesis. The requirement for transcription raises two fundamental cell biological questions: (1) how are signals transported from the synapse, where they are generated, to the nucleus, where they are converted into changes in gene expression? and (2) how are the products of gene expression targeted to alter structure and function at some but not all synapses made by a given neuron? The highly polarized morphology of neurons presents unique challenges to both of these processes (see Fig. 1). First, neurons elaborate dendrites and axons whose lengths often exceed the diameter of the cell body by orders of magnitude, indicating that signals generated at distal synapses must travel significant distances to reach the nucleus. Second, while each neuron has a single nucleus, it can form thousands of synaptic contacts, indicating that the products of gene expression must be targeted to alter the efficacy of stimulated synapses without affecting unstimulated synapses. In this chapter, we will discuss studies from our lab addressing both aspects of signaling between synapse and nucleus during learning-related synaptic plasticity. In studying signaling from the synapse to the nucleus, we focus on experiments demonstrating a role for the active, importin-mediated nuclear transport pathway. In studying the mechanisms whereby transcription produces synapse-specific forms of plasticity, we focus on synaptic mRNA localization and regulated translation.

There are a number of reasons to believe that basic research on the cell biological mechanisms of transcription-dependent synaptic plasticity will be highly relevant to understanding the pathophysiology of Alzheimer's disease (AD) and other neurodegenerative disorders. First and foremost, AD is a disease of memory and is thus likely to involve perturbations in some aspects of synaptic plasticity (Turner 2006). At a histological level, AD is characterized by a loss of synaptic density in the hippocampus and neocortex, and synaptic loss provides the best anatomical correlate of cognitive defects (Terry et al. 1991; Masliah et al. 2001). The amyloid precursor protein (APP), which plays a central role in AD pathogenesis, has been found to modify synaptic efficacy and structure and to affect synaptic plasticity (Venkitaramani et al. 2007; Shankar et al. 2007). These effects involve both intracellular and extracellular trafficking of proteolytic fragments of APP. The proteolytic cleavage of APP also generates an intracellular fragment, the AICD which, together with the

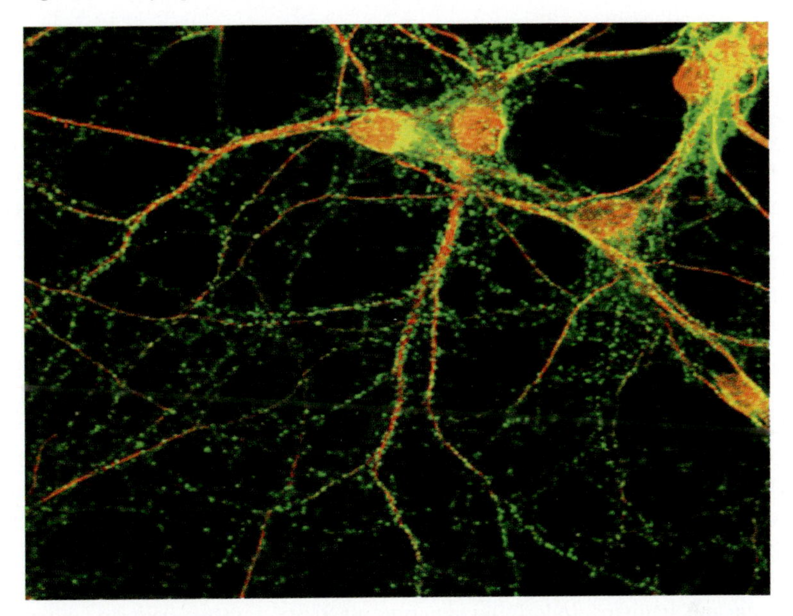

Fig. 1 Given the highly polarized morphology of neurons, signals received at distal synapses must travel significant distances to reach the cell body and the nucleus. Shown are cultured rat hippocampal neurons (14 days in vitro) stained for the synaptic marker PSD-95 (green) and the somatodendritic marker MAP-2 (red). Many of the PSD-95 immunoreactive synapses are present at significant distances from the cell soma, raising the question of how synaptically generated signals can be efficiently transmitted to the nucleus. Photomicrograph courtesy of Poon MM. From Heusner C and Martin KC, Signaling from the Synapse to the Nucleus, in Structural and Functional Organization of the Synapse, Michael Ehlers and Johannes Hell (eds), Springer, Berlin, Germany, in press, with permission

transcriptional regulator Fe65, is transported to the nucleus to initiate changes in transcription (Cao and Sudhof 2004; Chang et al. 2006). Finally, numerous studies have revealed defects in axonal transport during AD (and during other neurodegenerative diseases) that effectively disrupt bidirectional signaling between synapse and nucleus (Stokin and Goldstein 2006). Together, these findings suggest multiple ways in which research on the cell biology of signaling between synapse and nucleus during neuronal plasticity might shed light on the mechanisms underlying AD.

2 Model Systems for Studying the Cell Biology of Learning-Related Plasticity

We use two in vitro model systems to study learning-related synaptic plasticity: *Aplysia* sensory-motor synapses and cultured neurons from rodent hippocampus. The siphon sensory and gill motor neurons comprise a central component of the gill-withdrawal circuit in *Aplysia*, a defensive reflex in which the animal withdraws its gill following tactile stimulation of the siphon (Kandel 2001). This reflex undergoes sensitization, a nonassociative form of learning in which a mild electrical shock given to the tail of the animal leads to enhancement of the gill-withdrawal reflex. Sensitization occurs in short-term forms or in long-term forms, depending on the number of shocks the animal receives. Circuit-level studies have revealed that tail shock activates serotonergic interneurons that release serotonin onto the siphon sensory-gill motor neuron synapses, increasing synaptic efficacy and thereby producing sensitization. Most important for cell biological analyses, this circuit can be reconstituted in culture (Fig. 2), where one or two sensory neurons form monosynaptic connections with a single motor neuron. These synapses undergo both short- and long-lasting strengthening (short-term facilitation, STF; long-term facilitation, LTF) in response to direct application of serotonin (5HT; Kandel 2001). Thus, a single five-minute application of 5HT produces STF, whereas five spaced applications (5 min each) of 5HT produces LTF that persists at least 24 hours, requires new transcription and translation, and is accompanied by the growth of new synaptic connections. *Aplysia* neurons are large enough to allow imaging of the structural changes that accompany synaptic plasticity and microinjection of compounds (such as siRNAs, expression vectors and antibodies) to perturb, and thereby define, the molecular mechanisms underlying synaptic plasticity. Further, *Aplysia* neurons are robust enough to allow for repeated intracellular recordings over periods of at least 72 hours (Martin et al. 1997a). Studies using this system have identified a role for nuclear translocation of PKA and MAPK, for the induction of a cascade of gene expression that involves CREB-mediated transcription and leads to the production of immediate early and late effector genes that increase synaptic transmission

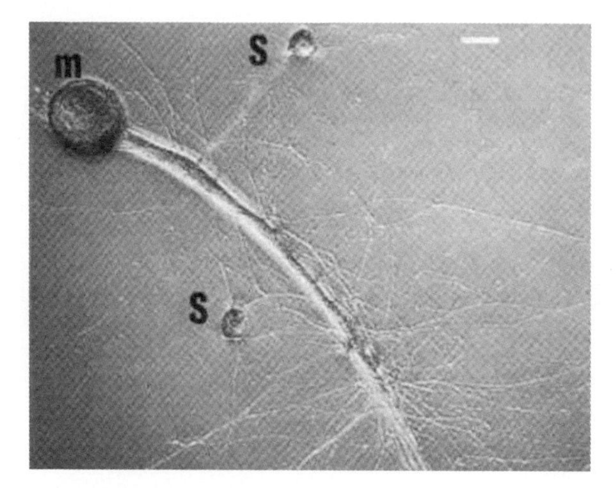

Fig. 2 *Aplysia* sensory neurons form synapses with motor neurons in culture. Shown are two sensory (S) neurons making synapses with a motor (M) neuron at 4 days in vitro. At this time, the excitatory postsynaptic potential formed between the sensory and motor neurons is stable and undergoes transient increases with a single application of serotonin (5HT) and long-lasting increases (over 24 hrs) with five spaced applications of 5HT. Scale bar = 50 μm

and promote the growth of new connections between sensory and motor neurons (Pittenger and Kandel 2003).

The mammalian hippocampus is critical for spatial and other forms of explicit memory and also undergoes shorter- and longer-lasting forms of synaptic strengthening (long-term potentiation, LTP). The long-lasting forms of LTP (L-LTP) share many mechanisms with LTF of *Aplysia* sensory-motor synapses. Specifically, L-LTP shows a requirement for nuclear translocation of PKA catalytic subunit and ERK 1/2 and activation of CRE-driven gene expression (Lonze and Ginty 2002). Numerous studies have revealed a role for synaptic mRNA localization and regulated translation in hippocampal neurons during learning-related plasticity. From an experimental perspective, the anatomy of the hippocampus allows stimulation and recording from each of the three types of excitatory synapses in the tri-synaptic circuit (the perforant, mossy fiber, and Schaeffer collateral pathways). Synapses are also formed by dissociated hippocampal neurons in culture, where the cells are more accessible to imaging and manipulations such as expression of recombinant proteins. As an experimental system, LTP of rodent hippocampus allows use of the extensive molecular knowledge and genetic manipulations available in the mouse.

3 Signaling From Synapse to Nucleus: How Does a Signal Get from a Distal Synaptic Compartment to the Nucleus to Initiate Changes in Gene Expression?

That soluble signals do indeed travel from distal sites of stimulation to the nucleus during plasticity is perhaps best illustrated by studies we have performed using an *Aplysia* culture system in which a single, bifurcated sensory neuron is cultured with two spatially separated motor neurons (Martin et al. 1997a; Fig. 3). In this culture system, application of 5HT to the connections made onto one of the motor neurons leads to LTF of that connection, and this LTF is dependent on transcription in the sensory neuron, whose nucleus is approximately 500 microns away from the site of stimulation (Martin et al. 1997a).

How might such signaling occur? Neurons are specialized for rapid signaling between compartments. Thus, depolarization at the synapse can spread passively or by action potentials to the cell soma, where voltage-dependent calcium channels can be activated, followed by rapid signaling to the nucleus. Since the endoplasmic reticulum (ER) is continuous with the nuclear envelope, and also extends out to distal synaptic sites, activation of IP3 and/or ryanodine receptors in the ER close to the synapse could produce regenerative calcium waves traveling from the synapse to the nucleus (Berridge 1998). Further, retrograde signaling has been shown to occur via signaling endosomes. Thus, for example, binding of NGF to the TrkA receptor leads to internalization of the ligand-bound receptor and trafficking of the complex to the cell body, where it can signal to the nucleus to produce changes in transcription. Finally, synaptic activity can activate soluble signaling molecules, such as the kinases protein kinase A (PKA) and ERK1/2, which travel through the

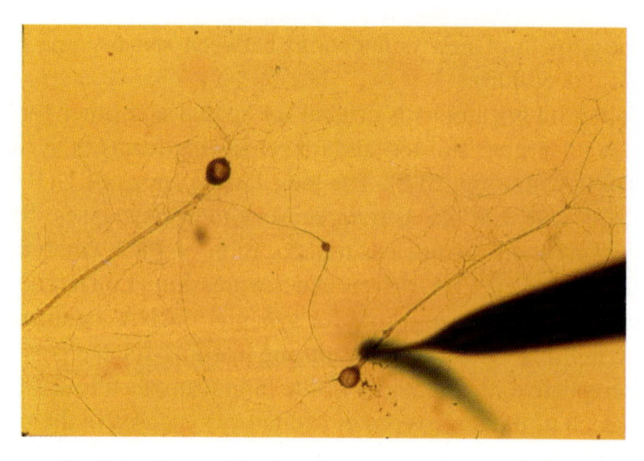

Fig. 3 Synapse-specific, long-term facilitation. A bifurcated sensory neuron forms synapses with two spatially separated motor neurons in culture (approximately 1 mm apart). Application of serotonin to the connections made onto one of the motor neurons produces long-term facilitation of that connection without altering synaptic efficacy at the opposite branch. This branch-specific, long-term facilitation requires transcription in the sensory neuron. Modified from Martin et al. (1997a)

neuronal process back to the cell body and ultimately into the nucleus. Transport of soluble molecules could occur by passive diffusion or by an active transport process.

During LTF of *Aplysia* sensory-motor synapses, 5HT does not depolarize sensory neurons; neither does it lead to increases in intracellular calcium in the sensory neuron (Blumenfeld et al. 1992; Eliot et al. 1993). Rather, the 5HT receptor is coupled to adenylate cyclase, leading to increases in cAMP and activation of PKA and MAP kinase (Goelet et al. 1986; Martin et al. 1997b). These soluble molecules translocate into the nucleus (Backsai et al. 1993; Martin et al. 1997b) and initiate a cascade of gene expression required for the persistent changes in synaptic efficacy and structure that mediate LTF. In considering how these synaptically generated signals might travel to the nucleus, we asked whether the importin family of nuclear transporters might play a role.

Importins (also known as karyopherins) have been characterized for their role in facilitating the transport of proteins across the nuclear pore and into the nucleus (Weis 1998). In the classical nuclear import pathway, proteins bearing nuclear localization signals (NLSs) within their primary sequence are recognized by one of a family of nuclear import adaptors called importin α. Importin α also binds to another nuclear transport factor, importin β1, which docks the complex at the nuclear pore and mediates its transport from cytoplasm into the nucleus. There are six importin α homologs in the human genome and five in the mouse genome, and there is evidence that these distinct isoforms (encoded by separate genes) show tissue specificity (Tsuji et al. 1997) and, further, specificity for the cargoes that they transport into the nucleus (Jans et al. 2000; Kohler et al. 1999).

We asked whether importins might be involved not only in facilitating transport across the nucleus but also in carrying karyophilic proteins from the synapse to the nucleus in neurons. In support of such a role, Richard Ambron and colleagues reported that a large cytoplasmic protein, human serum albumin (HSA), labeled with rhodamine, when microinjected into the growth cone of a cultured *Aplysia* neuron remained confined to the growth cone for at least 24 hrs (Ambron et al. 1992). When the rhodaminated HSA was coupled to an NLS and injected into a distal growth cone, however, it was rapidly and efficiently transported into the nucleus, indicating that the machinery for importin-mediated nuclear transport was present in distal processes. These earlier findings encouraged us to ask the following three questions: (1) do importins localize to synapses? (2) do importins travel from the synapse to the nucleus during transcription-dependent forms of plasticity? and (3) is importin-mediated transport required for long-lasting, transcription-dependent plasticity?

To address these questions, we first cloned an importin alpha isoform from *Aplysia*, ApImpα3, and generated antibodies against it. Immunocytochemistry with these antisera revealed that importin alpha was present in the cell soma, where it was concentrated at the nuclear membrane, and also in distal processes, where it was concentrated at synapses marked by synaptobrevin localization (Thompson et al. 2004). Immunoblotting of biochemical synaptoneurosome fractions confirmed the synaptic localization of importins. Together, these findings indicated that importins were appropriately localized to be involved in transporting signals from synapse to nucleus. We then asked whether stimuli that produce long-term, transcription-dependent plasticity triggered a relocalization of importins from synapse to nucleus. To address this question, we tagged ApImpα3 with GFP, expressed it in either the sensory or motor neuron (in a sensory-motor coculture), stimulated with either a single application of 5HT - to produce transcription-independent STF - or with five applications of 5HT, to produce transcription-dependent LTF, then imaged in real time by confocal microscopy. We found that ApImpα3 localization was not altered following a single application of 5HT but that five spaced applications of 5HT led to an accumulation of ApImpα-GFP in the nucleus of the sensory, but not the motor, neuron (Thompson et al. 2004 and unpublished data). Similar results were observed when sensory-motor cocultures were stimulated with forskolin, which also produces transcription-dependent LTF. These results indicate that stimuli that trigger LTF of *Aplysia* sensory-motor synapses trigger the transport of signals from the synapse to the nucleus. To determine whether or not importin-mediated transport was required for LTF, we injected anti-nuclear pore complex antibodies into sensory or motor neurons and determined their effects on basal synaptic transmission, STF and LTF. These antibodies have previously been shown to block active nuclear import without affecting passive diffusion through the nuclear pore. We found that the anti-nuclear pore antibodies blocked 5HT-induced LTF without affecting basal synaptic transmission or STF (Thompson et al. 2004 and unpublished data). This block was specific to microinjection into sensory neurons (unpublished data).

In complementary experiments performed in mouse hippocampal neurons (Thompson et al. 2004), we found that importins α1 and 2 were also localized to distal dendrites and axons, where they colocalized with synaptic markers. Biochemical fractionation revealed that both importin αs were present in postsynaptic density (PSD) fractions. We further found that activation of NMDA receptors led to a dramatic accumulation of importin α as well as importin β1 in nuclei. Induction of transcription-dependent LTP in acute hippocampal slices also triggered translocation of importins α1 and 2, and of importin β1, into the nucleus (Thompson et al. 2004). Together with our studies in *Aplysia*, these results provide strong evidence that importins localize to synapses and translocate to the nucleus, presumably bearing cargoes, following stimuli that trigger transcription in neurons.

Additional roles for importins have been identified during regeneration following neuronal injury. Michael Fainzilber and colleagues have shown that importin α3 and importin β1 are present in distal sciatic nerve axons and in the processes of cultured dorsal root ganglion (DRG) neurons (Hanz et al. 2003). Saturation of the active import pathway with excess NLS peptides delayed regenerative outgrowth in culture and inhibited the growth-enhancing effects of an in vivo conditioning lesion, consistent with importin-mediated signaling functioning to transport injury signals to the nucleus to initiate transcription-dependent regeneration (Hanz et al. 2003). These authors also reported that importin β1 mRNA was present in axons and that it was translated following injury. They subsequently defined a role for importin β1 in transporting phosphorylated MAP kinase from sites of injury to the nucleus (Perlson et al. 2005).

Studies have also uncovered a role for importins in mediating transport from growth cones and sites of synapse formation to the nucleus during development of the nervous system. Kumar et al. (2001) expressed dominant negative importin β in the *Drosophila* eye disc at a time when photoreceptors project their axons to the brain; they found that this prevented the axons from entering the optic stalk, leading to an extensive network of misguided axons. More recently, Larry Zipursky and colleagues (Ting et al. 2007) found that importin α3 mutations disrupted correct targeting of R7 photoreceptor axons in the *Drosophila* eyes. These reports indicate that axon guidance and cell adhesion in the eye rely on importin-mediated nuclear signaling.

The finding that importins play a role in signaling from the synapse to the nucleus raises a clear set of questions for future research. How are the importins localized to synapses, and how does synaptic stimulation regulate their retrograde nuclear transport? How do importins travel from distal synapses to the nucleus? The finding from Mike Fainzilber's research group that importin β1 interacts with the dynein motor protein via vimentin provides an important insight into the cell biological mechanisms that underlie importin-mediated signaling from synapse to nucleus and suggests that it involves dynein-dependent movement along microtubules. A third, central question concerns the identity of the synaptically localized cargoes that importins carry to the nucleus following stimulation. A recently published study identified Jacob, a caldendrin interacting protein, as a cargo of importin alpha that is translocated into the nucleus after NMDA receptor activation and is involved in the shutdown of CREB-dependent transcription (Dieterich et al. 2008). Bong-Kiun Kaang and colleagues have identified the *Aplysia* Cell Adhesion Molecule Associated Protein (CAMAP) as a cargo of ApImpα3 that is translocated from the sensory cell membrane to the nucleus during LTF (Lee et al. 2007). Our research group has undertaken both a candidate and an unbiased proteomic approach to identify potential importin cargoes at the synapse. As candidate cargoes, we have focused on transcriptional regulators whose activity is regulated by nucleocytoplasmic trafficking. For our proteomic approach, we are using coimmunoprecipitation with importin α antibodies from synaptoneurosome fractions, followed by mass spectrometry. Our hope is that we will be able to identify novel signaling molecules whose transport from synapse to nucleus is critical to transcription-dependent synaptic plasticity. Finally, perhaps the most critical question of all concerns identifying the type of stimuli that recruit importin-mediated signaling and the specific functions that importin-mediated synapse-nuclear signaling plays in neurons. Is importin-mediated nuclear import necessary for Hebbian, activity-dependent plasticity, for homeostatic plasticity, or for heterosynaptic forms of plasticity that require neuromodulatory inputs? Answers to these questions will likely emerge from studies that move the cell biological analyses in cultured neurons into an in vivo setting.

4 How can Transcription-Dependent Plasticity Occur in a Synapse-Specific Manner?

Given that the vertebrate brain contains approximately 10^{11} neurons but has 10^{14} synapses, information processing would clearly be greater if the unit of plasticity were the synapse rather than the nucleus. A number of studies have shown that subsets of synapses within an individual neuron can indeed undergo transcription-dependent plasticity. We demonstrated this at the level of a single cell using the preparation shown in Figure 3, in which a bifurcated sensory neuron contacted two spatially separated motor neurons. Five spaced applications of 5HT to one branch produced LTF of that branch without any change in synaptic strength at the opposite branch, and this branch-specific LTF depended on transcription in the sensory neuron.

In trying to determine the mechanisms whereby neurons could spatially restrict gene expression at the level of the synapse, we considered the possibility that it involved the local translation of mRNAs at stimulated synapses. We were influenced by a number of earlier findings. First, and most generally, mRNA localization and regulated translation have been shown to provide a means of spatially restricting gene expression in a number of asymmetric cells. Second, while most protein synthesis occurs in the cell bodies of neurons, polyribosomes, mRNAs and translation factors had been detected at synapses in hippocampal neurons (for review, see Steward and Schuman 2001). A more recent electron microscopic study has further shown that the number of polyribosomes at the base of spines of hippocampal CA1 pyramidal neurons triples two hours following LTP induction and that the PSDs of spines containing polyribosomes are significantly larger than those lacking polyribosomes (Ostroff et al. 2002). Third, in *Aplysia*, 5HT can elicit intermediate forms of facilitation (ITF) that require translation but not transcription (Ghirardi et al. 1995; Sutton et al. 2001), indicating that 5HT can modulate the translational machinery independent of its effects on transcription.

We tested the requirement for local translation by perfusing membrane permeant translational inhibitors either at the branch receiving the 5HT or at the opposite branch, and we found that inhibition of protein synthesis at the site receiving the 5HT completely blocked branch-specific LTF (Martin et al. 1997a). By culturing sensory neurons and removing their cell bodies (*Aplysia* neurons will survive for days without a cell body), we generated a preparation of pure sensory neurites. Metabolic labeling of this preparation showed that the neurites are capable of translation and that 5HT dramatically stimulates this translation (Martin et al. 1997a).

To determine the identify of the mRNAs that localize to the neurite and to the synapse, mRNAs that we considered likely to be important to synapse-specific plasticity, we made a cDNA library from the isolated sensory neurites (Moccia et al. 2003). Sequencing of this library revealed that it contained a surprisingly large number of transcripts: nearly 250 total distinct mRNAs, 100 of which constituted approximately 70% of the library. The library was a 3'EST library and was composed largely of sequences in the 3'UTRs of mRNAs. Given that the *Aplysia* genome has not yet been sequenced, we could not determine the identity of all of the mRNAs. We performed full-length cloning of 20 mRNAs and found that they were enriched in mRNAs encoding cytoskeletal elements and components of the translational machinery. In situ hybridization confirmed that all were present in distal neurites of *Aplysia* sensory neurons. We were further able to show that 5HT increased the translation of three of the transcripts, those encoding T1 α-tubulin, β thymosin and the sensory cell-specific mRNA sensorin (Moccia et al. 2003 and unpublished data).

We undertook a similar effort to identify dendritically localized mRNAs in rodent hippocampal neurons. To do this, we cultured neurons on filters containing 3-micron pores that confined the cell bodies to the top surface but allowed the neuronal axons and dendrites (and glial processes) to penetrate through and grow along the bottom surface (Poon et al. 2006). We mechanically isolated the bottom, process-containing surface and used this to generate probes for microarray analysis. Using this approach, we again identified a large number ($\sim$150) of mRNAs. In situ hybridization of 20 of these transcripts in cultured rat hippocampal neurons revealed that all were present in dendrites (Poon et al. 2006). Intriguingly, the localized mRNAs were enriched for mRNAs encoding components of the translational machinery.

The finding that many localized mRNAs, both in *Aplysia* and in rodent hippocampus, encoded molecules involved in translation led us to hypothesize that one of the functions of local translation might be to generate "translational sinks" localized sites of increased translation (Moccia et al. 2003). This would provide a mechanism of integrating the requirement for transcription with the finding that synapse-specific plasticity requires local translation. The idea is that, if local synaptic stimulation increased the translational capacity of the stimulated synapses and also recruited transcription in the nucleus, the products of gene expression would be preferentially translated at stimulated synapses. Notably, this idea is consistent with work from Kristen Harris and colleagues showing that those synapses that have polyribosomes

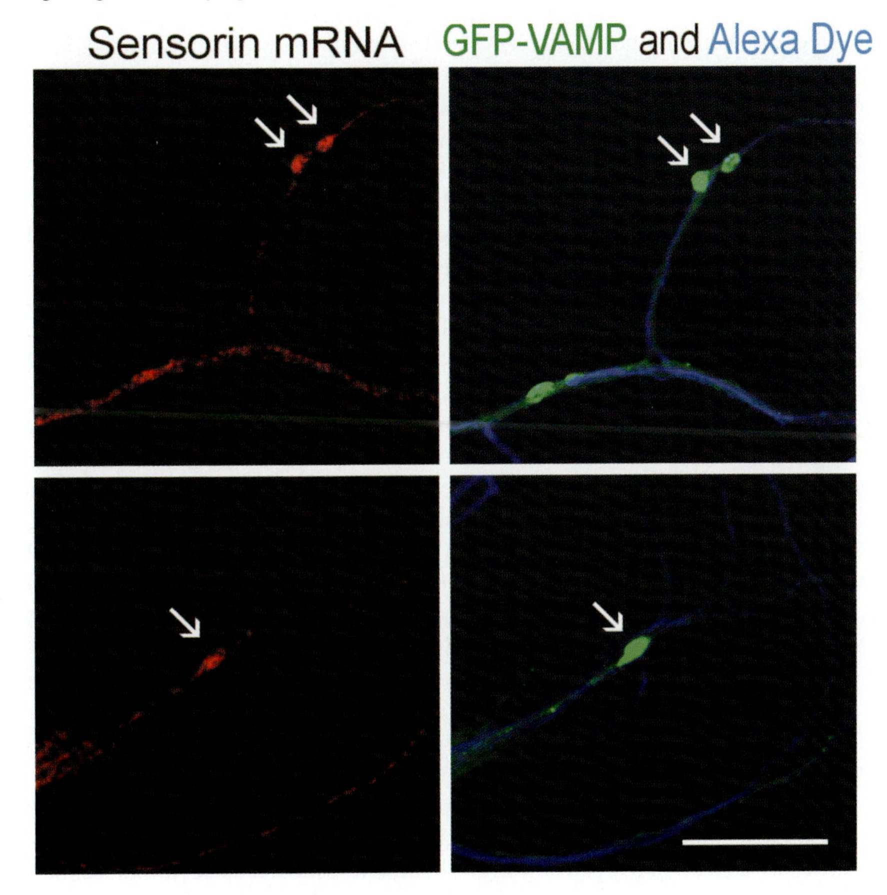

Fig. 4 Sensorin mRNA, detected by in situ hybridization (red), is expressed in sensory (and not motor) neurons and concentrates at sites of contact with the motor neuron. These sites colocalize with the synaptic marker GFP-VAMP, which is expressed in the sensory neuron. The motor neuron is labeled with Alexa-fluor 633, in blue, and the sensory neuron grows along the motor neuron to form synapses (marked by the green VAMP signal). Scale bar = 10 μm. Modified from Lyles et al. (2006)

associated with them following LTP induction in rat hippocampus have an increased size of their PSD (Ostroff et al. 2002).

We have used the *Aplysia* culture system to begin to study the function of localized mRNAs, since it allows us to visualize and manipulate mRNA localization at the level of individual neurons. As one example, in situ hybridization of the mRNA encoding the neuropeptide sensorin revealed that it localized diffusely to the neurites of isolated sensory neurons (which do not form synapses with themselves or with each other), but concentrated at sites of synaptic contact in sensory neurons forming synapses with motor neurons (Lyles et al. 2006). The protein showed a similar pattern of relocalization, suggesting that the mRNA was indeed locally translated into protein. We used RNAi to selectively knock down sensorin mRNA and analyzed the effect of localized translation on synapse formation by looking at time points at which the mRNA was degraded but the concentration of sensorin protein was not reduced. These experiments revealed that translation of sensorin mRNA was specifically required for the formation and/or stabilization of synapses between sensory and motor neurons.

Current efforts are aimed at using novel methods to visualize both mRNA localization and localized translation in living neurons. We are using a recently described method for visualizing mRNA trafficking in *Aplysia* neurons (Daigle and Ellenberg 2007). This system includes two basic components: 1) GFP fused to an arginine-rich 22-amino acid peptide derived from the phage λN protein, λN22 and 2) a unique minimal RNA sequence (a 15 nt hairpin RNA structure called box B) to which λN22 binds with high affinity (Kd of 22nM). Previously used for biochemical purification, Daigle and Ellenberg adapted the λN-boxB interaction to visualize mRNA trafficking in live cells. The system for mRNA visualization consists of two plasmids: one plasmid, called λN22-3mEGFP-M9, encodes four copies of λN fused to three copies of GFP and to the M9 nuclear localization signal (NLS); the second plasmid, called RNA-4boxB, encodes four copies of the box B λN22 binding site engineered into the 3'UTR of an mRNA of interest. When the λN22-3mEGFP-M9 is expressed alone, it is sequestered in the nucleus. When it is expressed with the RNA-4boxB, λN22-3mEGFP-M9 binds to the 4boxB sites and is transported out of the nucleus with the RNA into the cytoplasm, where its localization can be followed in real-time by GFP fluorescence. Using this system, we can address the question of whether mRNAs target to specific synapses during synapse formation and synaptic plasticity.

To visualize local translation in real-time, we are using reporter constructs encoding photoconvertible fluorescent proteins. Fusing the reporter constructs to the untranslated regions of localized mRNAs allows us to map the sequences that mediate mRNA localization (by performing in situ hybridization to detect the subcellular localization of the reporter RNA). The advantage of using a photoconvertible fluorescent protein for these experiments is that we can photoconvert the fluorophore from its native conformation (green) to the photoconverted conformation (red), and then detect any new translation as green, unconverted signal. These experiments should allow us to detect new translation at the level of individual synapses during both synapse formation and synaptic plasticity.

5 Conclusions

The requirement for transcription during long-lasting, learning-related synaptic plasticity raises two fundamental cell biological questions: (1) how are signals transported from distal synaptic sites to the nucleus to initiate changes in gene expression and (2) how are the products of gene expression targeted to alter efficacy at some but not all synapses made by a given neuron? Using two model systems of learning-related synaptic plasticity, we have discovered a role for importin-mediated trafficking in carrying synaptically generated signals to the nucleus during long-term plasticity. Importins are present at synapses and translocate to the nucleus following stimuli that induce transcription. In *Aplysia* sensory-motor synapses, importin-mediated nuclear transport is required for LTF but not STF. Current efforts are aimed at identifying the synaptically localized cargoes that importins carry to the nucleus, at elucidating the types of stimuli that recruit importin mediated signaling from synapse to nucleus, and at characterizing the cell biological pathways whereby importins travel retrogradely to the nucleus. In terms of understanding how a transcription-dependent process could occur in a synapse-specific manner, we have focused on a role for localizing mRNAs and regulating their translation by synaptic stimulation. We have identified hundreds of localized mRNAs in both *Aplysia* sensory and rodent hippocampal neurons, and we have confirmed the localization of many of these by in situ hybridization. Current efforts are aimed at visualizing mRNA localization and regulated translation in living neurons and at using RNA interference to elucidate the function of local translation during synapse formation and synaptic plasticity.

References

Alberini CM (1999) Genes to remember. J Exp Biol 202: 2887–2891

Ambron RT, Schmied R, Huang CC, Smedman M (1992) A signal sequence mediates the retrograde transport of proteins from the axon periphery to the cell body and then into the nucleus. J Neurosci 12: 2813–2818

Bacskai BJ, Hochner B, Mahaut-Smith M, Adams SR, Kaang BK, Kandel ER, Tsien RY (1993) Spatially resolved dynamics of cAMP and protein kinase A subunits in *aplysia* sensory beurons. Science 260: 222–226

Berridge MJ (1998) Neuronal calcium signaling. Neuron 21: 13–26

Blumenfeld H, Zablow L, Sabatini B (1992) Evaluation of cellular mechanisms for modulation of calcium transients using a mathematical model of fura-2 Ca2+ imaging in *aplysia* sensory neurons. Biophys J 63: 1146–1164

Bruel-Jungerman E, Davis S, Laroche S (2007) Brain plasticity mechanisms and memory: a party of four. Neuroscientist 13: 492–505

Cao X, Sudhof TC (2004) Dissection of amyloid-beta precursor protein-dependent transcriptional transactivation. J Biol Chem 279: 24601–24611

Chang KA, Kim HS, Ha TY, Ha JW, Shin KY, Jeong YH, Lee JP, Park CH, Kim S, Baik TK, Suh YH (2006) Phosphorylation of amyloid precursor protein (APP) at Thr668 regulates the nuclear translocation of the APP intracellular domain and induces neurodegeneration. Mol Cell Biol 26: 4327–4338

Daigle N, Ellenberg J (2007) LambdaN-GFP: an RNA reporter system for live-cell imaging. Nature Methods 4: 633–636

Dieterich DC, Karpova A, Mikhaylova M, Zdobnova I, Konig I, Landwehr M, Kreutz M, Smalia KH, Richter K, Landgraf P, Reissner C, Boeckers TM, Zuschratter W, Spilker C, Seidenbecher CI, Garner CC, Gundelfinger ED (2008) Caldendrin-Jacob: a protein liaison that couples NMDA receptor signaling to the nucleus. PLoS Biol 6: e34.

Dynes JL, Steward O (2007) Dynamics of bidirectional transport of arc mRNA in neuronal dendrites. J Comp Neurol 500: 433–447

Eliot LS, Kandel ER, Siegelbaum SA, Blumenfeld H (1993) Imaging terminals of *aplysia* sensory neurons demonstrates role of enhanced Ca2+ influx in presynaptic facilitation. Nature 361: 634–637

Ghirardi M, Montarolo PG, Kandel ER (1995) A novel intermediate stage in the transition between short- and long-term facilitation in the sensory to motor neuron synapse of *aplysia*. Neuron 14: 413–420

Goelet P, Castellucci VF, Schacher S, Kandel ER (1986) The long and the short of long-term memory–a molecular framework. Nature 322: 419–422

Hanz S, Perlson E, Willis D, Zheng JQ, Massarwa R, Huerta JJ, Koltzenburg M, Kohler M, van-Minnen J, Twiss JL, Fainzilber M (2003) Axoplasmic importins enable retrograde injury signaling in lesioned nerve. Neuron 40: 1095–1104

Jans DA, Xiao CY, Lam MH (2000) Nuclear targeting signal recognition: a key control point in nuclear transport? Bioessays 22: 532–544

Kandel ER (2004) The molecular biology of memory storage: a dialog between genes and synapses. Biosci Rep 24: 475–522

Kim SJ, Linden DJ (2007) Ubiquitous plasticity and memory storage. Neuron 56: 582–592

Kohler M, Speck C, Christiansen M, Bischoff FR, Prehn S, Haller H, Gorlich D, Hartmann E (1999) Evidence for distinct substrate specificities of importin alpha family members in nuclear protein import. Mol Cell Biol 19: 7782–7791

Kumar JP, Wilkie GS, Tekotte H, Moses K and Davis I (2001) Perturbing nuclear transport in drosophila eye imaginal discs causes specific cell adhesion and axon guidance defects. Dev Biol 240: 315–325

Lee SH, Lim CS, Park H, Lee JA, Han JH, Kim H, Cheang YH, Lee SH, Lee YS, Ko HG, Jang DH, Kim H, Miniaci MC, Bartsch D, Kim E, Bailey CH, Kandel ER, Kaang BK (2007) Nuclear translocation of cam-associated protein activates transcription for long-term facilitation in *aplysia*. Cell 129: 801–812

Lonze BE, Ginty DD (2002) Neuron 35:605–623

Lyles V, Zhao Y, Martin KC (2006) Synapse formation and mRNA localization in cultured *aplysia* neurons. Neuron 49: 349–356

Martin KC, Casadio A, Zhu H, Yaping E, Rose JC, Chen M, Bailey CH, Kandel ER (1997a) Synapse-specific, long-term facilitation of *aplysia* sensory to motor synapses: a function for local protein synthesis in memory storage. Cell 91: 927–938

Martin KC, Michael D, Rose JC, Barad M, Casadio A, Zhu H, Kandel ER (1997b) Map kinase translocates into the nucleus of the presynaptic cell and is required for long-term facilitation in *aplysia*. Neuron 18: 899–912

Masliah E, Mallory M, Alford M, DeTeresa R, Hansen LA, McKeel DW Jr, Morris JC (2001) Altered expression of synaptic proteins occurs early during progression of Alzheimer's Disease. Neurology 56: 127–129

Moccia R, Chen D, Lyles V, Kapuya E, E Y, Kalachikov S, Spahn CM, Frank J, Kandel ER, Barad M, Martin KC (2003) An unbiased cDNA library prepared from isolated *aplysia* sensory neuron processes is enriched for cytoskeletal and translational mRNA. J Neurosci 23: 9409–9417

Neves G, Cooke SF, Bliss TV (2008) Synaptic plasticity, memory and the hippocampus: a neural network approach to causality. Nature Rev Neurosci 9: 65–75

Ostroff LE, Fiala JC, Allwardt B, Harris KM (2002) Polyribosomes redistribute from dendritic shafts into spines with enlarged synapses during LTP in developing rat hippocampal slices. Neuron 35: 535–545

Perlson E, Hanz S, Ben-Yaakov K, Segal-Ruder Y, Seger R, Fainzilber M (2005) Vimentin-dependent spatial translocation of an activated MAP Kinase in injured nerve. Neuron 45: 715–726

Pittenger C, Kandel ER (2003) In search of general mechanisms for long-lasting plasticity: *aplysia* and the hippocampus. Philos Trans R Soc Lond B Biol Sci 358: 757–763

Poon MM, Choi SH, Jamieson CA, Geschwind DH, Martin KC (2006) Identification of process-localized mRNAs from cultured rodent hippocampal neurons. Neurosci 26: 13390–13399

Shankar GM, Bloodgood BL, Townsend M, Walsh DM, Selkoe DJ, Sabatini BL J (2007) J Neurosci. 27:2866–2875

Steward O, Schuman EM (2001) Protein synthesis at synaptic sites on dendrites. Annu Rev Neurosci 24: 299–325

Stokin GB, Goldstein LS (2006) Axonal transport and Alzheimer's disease. Annu Rev Biochem 75: 607–627

Sutton MA, Masters SE, Bagnall MW, Carew TJ (2001) Molecular mechanisms underlying a unique intermediate phase of memory in *aplysia*. Neuron 31: 143–154

Terry RD, Masliah E, Salmon DP, Butters N, DeTeresa R, Hill R, Hansen LA, Katzman R (1991) Physical basis of cognitive alterations in Alzheimer's Disease: synapse loss is the major correlate of cognitive impairment. Ann Neurol 30: 572–580

Thompson KR, Otis KO, Chen DY, Zhao Y, O'Dell TJ, Martin KC (2004) Synapse to nucleus signaling during long-term synaptic plasticity; a role for the classical active nuclear import pathway. Neuron 44: 997–1009

Ting CY, Herman T, Yonekura S, Gao S, Wang J, Serpe M, O'Connor MB, Zipursky SL, Lee CH (2007) Tiling of R7 axons in the drosophila visual system is mediated both by transduction of an activin signal to the nucleus and by mutual repulsion. Neuron 56: 793–806

Tsuji L, Takumi T, Imamoto N, Yoneda Y (1997) Identification of novel homologues of mouse importin alpha, the alpha subunit of the nuclear pore-targeting complex, and their tissue-specific expression. FEBS Lett 416: 30–34

Turner RS (2006) Alzheimer's disease. Semin Neurol 26: 499–506

Venkitaramani DV, Chin J, Netzer WJ, Gouras GK, Lesne S, Malinow R, Lombroso PJ (2007) Beta-amyloid modulation of synaptic transmission and plasticity. J Neurosci 27: 11832–11837

Weis K (1998) Importins and exportins: how to get in and out of the nucleus. Trends Biochem Sci 23: 185–189

Axonal Transport of Neurotrophic Signals: An Achilles' Heel for Neurodegeneration?

Ahmad Salehi(✉), Chengbiao Wu, Ke Zhan, and William C. Mobley

Abstract The most effective treatments for neurodegenerative disorders, including Alzheimer's disease, will come from studies of the pathogenesis of age-related cognitive failure and understanding of the underlying mechanisms. Given the marked similarities in pathological and clinical phenotypes between Alzheimer's disease and Down syndrome, studies of the pathogenesis of one can be expected to complement and support those in the other. Alzheimer's disease and Down syndrome are characterized by dysfunction and loss of several biochemically and anatomically defined neuronal populations. The pathological involvement of hippocampus, in particular, is an early feature of both disorders, as is the degeneration of neurons whose axons innervate this region. Long, thin and poorly myelinated axons project from a number of subcortical and brain stem nuclei to modulate hippocampally mediated cognitive functions. In studies on mouse models of Down's syndrome, we uncovered evidence for the involvement of a particular neuronal population heavily innervating the hippocampus. In an extensive series of experiments, we found evidence that failed retrograde transport of nerve growth factor signaling in cholinergic neurons of the basal forebrain is linked to their vulnerability and that these changes are caused by increased gene dose and overexpression of the gene for amyloid precursor protein. These findings raise the possibility that intracellular trafficking defects created by changes in amyloid precursor protein expression or processing make an important contribution to pathogenesis and set the stage for studies to explore the molecular mechanisms of degeneration of cholinergic neurons and to define new therapeutic targets for these neurons. An important unanswered question is whether or not similar mechanisms operate within other vulnerable populations, innervating hippocampus to cause de-afferentation and dysfunction of this critical brain region.

A. Salehi
Stanford University School of Medicine, Dept of Neurology, CA 94305, Stanford, USA
E-mail: asalehi@stanford.edu

P. St. George-Hyslop et al. (eds.) *Intracellular Traffic and Neurodegenerative Disorders*, Research and Perspectives in Alzheimer's Disease,

1 Introduction

Age-related neurodegenerative disorders are characterized by degeneration of specific neuronal populations, i.e., the selective involvement of certain neurons. Before cell death, degenerating neurons usually show shrinkage, reduction or loss of markers, as well as changes in the morphology of dendrites (Morrison and Hof 2002; Belichenko et al. 2004). Axonal involvement is also often prominent. Indeed, it features (1) synaptic dysfunction and loss; (2) axonal pathology, often severe; and (3) the presence of proteinaceous inclusions composed of misfolded proteins. All of these markers may significantly predate neuronal atrophy, degeneration and death. In light of this chronology, it is important to explore the changes that occur early in the course of neurodegeneration and to decipher their molecular pathogenesis. Important additional sources of insight come from studies of the genetics of neurodegeneration and from molecular and cellular studies to evaluate the effects of the protein products of the responsible genes. Herein, we explore the hypothesis that selective vulnerability is engendered, at least in part, in the failure of axons to transport neurotrophic signals from axons in targets to the cell bodies of responsive neurons. An emerging story that links increased gene dose for amyloid precursor protein (*APP*) to axonal dysfunction and age-related degeneration in Down syndrome (DS) may provide unique insights in the pathogenesis of Alzheimer's disease (AD).

1.1 The Axon as a Focus of Attention

We have been interested in the genesis of synaptic and axonal pathology in neurodegeneration. The axon plays a unique and critical role in the biology of the neuron. It represents the conduit for carrying anterogradely most if not all of the materials needed to provide axon terminals with the molecular machinery needed to carry out neurotransmission. In addition, it is the route by which retrograde transport carries synaptic proteins for degradation. Most relevant to the current work, it is the link by which neurotrophic signals produced in postsynaptic target neurons are sent to cell bodies to instruct the neuronal nucleus to support continued maintenance of synaptic contacts and, thereby, the integrity of neuronal circuits. Remarkably, the axon carries out these functions with space and time constraints that are quite extraordinary. The length of an axon may be more than 1,000 times the diameter of its cell body. It carries traffic over these long distances using a variety of motor proteins and does so at speeds in the range of 1 to several μm/second (Howe and Mobley 2005).

1.2 Axons Carry Neurotrophic Signals

Retrograde trophic signaling is essential for the survival and differentiation of developing neurons and for the maintenance of function of mature neurons (Sofroniew et al. 2001). Recent studies in this and other laboratories have defined signaling endosomes as important organelles for retrogradely transporting the neurotrophic signals of nerve growth factor (NGF) and other neurotrophins (Heerssen and Segal 2002; Ginty and Segal 2002; Delcroix et al. 2004; Howe and Mobley 2005). The "signaling endosome hypothesis" speaks to the mechanisms by which trophic signals are produced within and carried by this organelle. Neurotrophic factors released from cells in the target of innervation diffuse to, bind, and activate their specific receptors, and the complex thus formed is internalized. Interestingly, the endosome that results bears on its surface most or all of the signaling proteins that are needed for executing the activation of the mitogen-activated protein kinases (MAPKs, i.e., Erk1/2, Erk5), PI3k/Akt, and possibly the phospholipase C-γ (PLC-γ) pathways (Heerssen and Segal 2002; Ginty and Segal 2002; Wu et al. 2007). Signaling endosomes are then transported via dynein-based transport along microtubules to the cell body. There is compelling evidence that this endosome signals during transit as well as upon arrival in the soma. What significance can be attached to signaling-in-transit is unknown, but one can readily imagine that such signals could be used to inform that axon of the status of its target.

1.3 Scaling Axonal Traffic to Appreciate the Dynamics

It is perhaps useful to scale these measures to demonstrate that movement is long-range, rapid and vulnerable to failure. If we use its diameter of 100 nm to scale an endosome scaling to the size of an automobile, it would travel in a tube of about 100 ft in diameter at a rate of 80 m/sec or 288 km/hr for a distance of about ~3,500 km over 12 hours. At this rate, it could travel from San Francisco to New York City in about 17 hours. It would do so on an undulating roadway and in the congested confines of a tube that contains a number of both relatively stationary (i.e., microtubules and assembled neurofilaments) and mobile elements. It would pass or be passed by other mobile elements moving retrogradely and would encounter oncoming anterograde traffic; the speed of convergence would be almost 600 km/hr. Local changes in the integrity of the roadway would be present. Certainly, inclusions could readily be envisioned to disrupt or stall traffic. Tangles could nearly occlude the tube. Moreover, the transport would be continually dependent not on an internal fuel supply but on power plants (i.e., mitochondria) distributed along the axon. The tube itself would be drawing on this same source of energy as it carried electrical signals (i.e., action potentials) at a speed that would be more than 1 million-fold faster than the speed of the endosome. Taken together, these findings indicate that the retrograde traffic of trophic signals in axons is confronted by substantial physiological barriers on a dynamic milieu.

1.4 Axonal Dysfunction and Neurodegeneration

Several recent observations link genetic mutations to neurodegenerative disorders. The following are caused by mutations in proteins that regulate axonal transport or that act to disrupt transport as mutants. *APP* mutations and duplication have been linked to familial AD (FAD). Studies from our laboratory and others have shown that early endosomes (EEs) are abnormally enlarged and contain App and its C-terminal fragments (Salehi et al. 2006). Very recently, two sets of single nucleotide polymorphism (SNPs) in *SORL1* gene were linked to familial as well as sporadic forms of AD (Rogaeva et al. 2007). SORL1 is a glycoprotein receptor that is believed to play a major role in endosomal transport in neurons. Furthermore, it appears that SORL1 plays a significant role in APP metabolism and trafficking through the endocytic compartment and trans Golgi network (Schmidt et al. 2007; see Table 1 for other examples). Studies showing that alterations in axonal structure or function are early, and significant markers of pathogenesis would bring new insights to bear on molecular mechanisms and could provide novel methods and tools for the early diagnosis and treatment of these disorders.

1.5 The Hippocampus: Evidence for De-afferentation in AD

The hippocampal formation plays a crucial role in a variety of higher cognitive functions, including learning and memory. Proper function depends on integrity of intrahippocampal circuits as well as projections from cortical and subcortical regions. Its internal circuit structure includes (1) the dentate gyrus (DG), whose activity is regulated by local networks of interneurons; (2) the CA3 region, whose pyramidal neurons receive excitatory input from the DG; and (3) the CA1 region, whose pyramidal neurons receive excitatory input from CA3 and send inputs to the subiculum through the stratum oriens (Fig. 1). The main cortical input to the hippocampus is the perforant pathway, whose axons originate in the entorhinal cortex (EC layers II and III in the rat) and whose principal excitatory input is delivered to the DG. The subcortical regions that send extensive projections to the hippocampus in rodents include cholinergic neuron in the basal forebrain (BCFN; the medial septal nucleus and diagonal bands, MSDB), noradrenergic neurons in locus coeruleus (LC), serotoninergic neurons of raphe nuclei (RN), and neurons of the supramamillary area (SUMA). These relatively large but numerically scarce neurons project extensively to specific groups of neurons in the hippocampus. For instance, the MSDB complex sends large projections from cholinergic as well as GABA-ergic neurons to the hippocampus. In the hippocampus, the supragranular region, $\sim 1/4 - 1/3$ of the molecular layer in the immediate vicinity of the DG cell layer, receives the densest cholinergic projections making mostly symmetrical synapses with the dendrites of DG cells. The majority of GABA-ergic terminals in the DG end in the subgranular layer (e.g., GABA-ergic chandelier and basket cells) and the polymorphic layer of the DG. LC is the sole source of noradrenergic

Table 1

Gene	**Disease**	**Sign & Symptoms**	**Role of the Encoded Protein in Transport**
APP (1)	Alzheimer's disease	Dementia	TrkA-NGF signaling (2)
SORL (3)	Alzheimer's disease	Dementia	Endosomal transport, App metabolism and transport (4)
ALS2 (5)	ALS2	Muscular atrophy	Rab5 activation, Endosomal trafficking (6)
p150(glued)(7)	dSBMA	Muscular atrophy	Vesicle transport (8)
TAU (9)	FTDP-17	Behavioral, motor and cognitive dysfunction	A cytoskeletal protein, Interaction with p150(glued) (10)
HTT (11)	Huntington's disease	Motor dysfunction and cognitive impairment	Kinesin-mediated mitochondria transport (12)
HSP27 (13)	CMT2	Motor and sensory neuropathy	Actin stabilization (14)
KIFIB (15)	CMT	Motor and sensory neuropathy	Motor protein (16)
SNCA (17)	Parkinson's disease	Motor and cognitive dysfunction	Vesicle transport (18)
SODI (19)	ALS	Muscular atrophy	Interaction with Dynein (20)
GAN (21)	GAN	Sensory motor neuropathy	Interacting with cytoskeletal proteins (22)
RAB7 (23)	CMT	Motor and sensory neuropathy	Vesicle transport (24)

APP; Amyloid precursor protein, ALS, Amyotrophic lateral sclerosis, DSBMA, distal spinal and bulbar muscular atrophy. *SORL1*; neuronal sortilin-related receptor, *HTT*, Huntingtin, ARSCCS; autosomal recessive spastic ataxia of Charlevoix-Saguenay, CMT; Charcot-Marie-Tooth disease. *HSP27*; heat shock protein 27, DHMN: Distal hereditary motor neuropathies SPG13. Hereditary spastic paraplegia. *SNCA*, Synuclein alpha, GAN; giant axonal neuropathy.
(1) Goate et al., 1991. (2) Salehi et al., 2006. (3) Rogaeva et al., 2007. (4) Offe et al., 2006. (5) Yang et al., 2001. (6) Kunita et al., 2007. (7) Puls et al., 2005. (8) Laird et al., 2008. (9) Hutton et al., 1998. 10) Magnani et al., 2007. (11) The Huntington's Disease Collaborative Research Group (1993). (12) Orr et al., 2008. (13) Evgrafov et al., 2004. (14) Lavioe et al., 1995. (15) Kijima et al., 2005. (16) Hirokawa and Takemura, 2003. (17) Polymeropoulos et al., 1997. (18) Gitler et al., 2008. (19) Rosen et al., 1993. (20) Ström et al., 2008. (21) Bomont et al., 2000. (22) Yang et al., 2007. (23) Meggouh et al., 2006. (24) Ng and Tang, 2008.

terminals in the hippocampus. These terminals end mostly in the DG and stratum lucidum of the CA3 region. The serotoninergic innervation of the hippocampus originates mostly from dorsal raphe (DR) and median raphe nucleus (MRN). Projections from the RN in the DG terminate in the subgranular area and in the polymorphic layer, making synapses with GABAergic neurons. The calretinin-positive neurons of the SUMA send major projections either directly or indirectly through the MSDB to the DG of the hippocampus. The majority of these neurons terminate in the supragranular region of the molecular layer in the immediate vicinity of DG cell layer, making synapses with the primary dendrites of DG cells. Furthermore, the pyramidal layer of the CA2 area also receives heavy innervations from SUMA neurons. In humans, the SUMA together with tuberomamillary nuclei constitutes the histaminergic tuberomamillary nuclear complex.

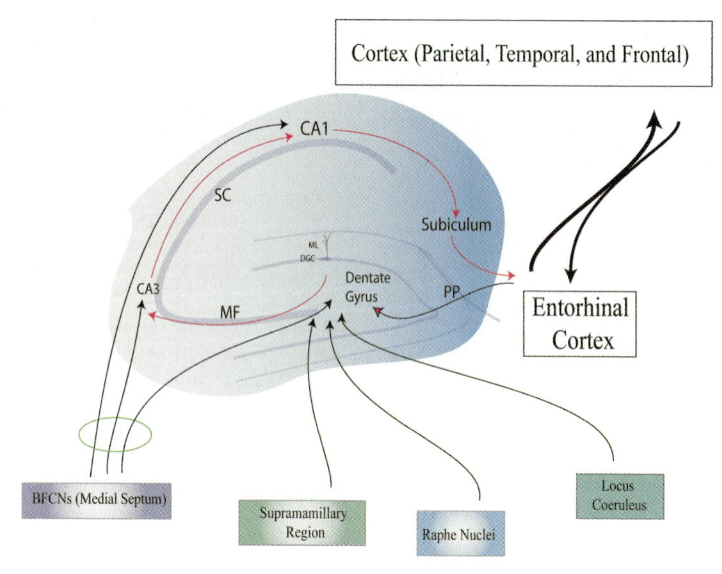

Fig. 1 Schematic representation of the sagittal view of the mouse hippocampus with its main afferents from (MSN and DB) MSDB complex, LC, SUMA, and RN

The integrity of the major inputs to the hippocampus plays a crucial role in its normal physiology. It has been shown that lesions or inactivation of SUMA (Shahidi et al. 2004), septum (Moreau et al. 2008), MRN (Borelli et al. 2005), and LC (Compton et al. 1995) in rodents lead to impaired learning and memory.

The hippocampus is an early site of pathology in AD. Especially noteworthy is the presence of neurofibrillary tangles. Indeed, it appears that only pathology in the entorhinal cortex precedes that for hippocampus. Interestingly, extrahippocampal regions undergo extensive degeneration in the course of AD (Braak et al. 1999). Thus, in addition to entorhinal cortex, the nucleus basalis of Meynert, LC, RN and neurons in the TM show extensive atrophy and degeneration and AD pathology. Thus, the systems affected include, but are not limited to, specific sets of cholinergic, serotoninergic, noreadrenergic, histaminergic, and dopaminergic neurons.

As yet undetermined is whether or not a unifying hypothesis can be proposed to explain degeneration of these morphologically and functionally related populations. Conceivably, simply their projection to the markedly affected hippocampus would be enough to predispose them to degeneration. Synaptic dysfunction and disconnection can readily be envisioned to suffice. But it would be interesting and potentially important to explore the possibility that other events preceding synaptic dysfunction play a role. In a search for features common to neurons whose axons extend to hippocampus that are vulnerable in AD, we note that all these populations are responsive and retrogradely transport neurotrophins (see Mufson et al 1999; Celada et al. 1996). For instance, in rodents, it has been shown that BFCNs and SUMA retrogradely transport NGF, whereas LC, RN, and EC transport BDNF (Mufson et al. 1999). It was noted earlier that all neurons with thin, poorly myelinated axons that project for relatively large distances to their targets are prone to degeneration. Indeed sensory primary and motor primary fields that are heavily myelinated are scarcely affected by plaque and tangles. However, the entorhinal and hippocampal regions, which are poorly myelinated, are generally heavily affected in AD (Braak et al. 1999). Though hardly a unique set of relationships, as many other populations with thin axons are dependent on neurotrophins, the convergence of these observations with the anatomy of neurodegeneration in AD point to the possibility that failed neurotrophic signaling in the axons of afferent populations may contribute to their degeneration.

1.6 Degeneration of BFCNs in AD: Evidence for Failed NGF Transport and Signaling

Due to the facts that BFCNs invariably degenerate in the course of AD, leading to cholinergic de-afferentation of the hippocampus, and that NGF signaling plays a significant role in phenotypic maintenance of these neurons, much attention has been devoted to studying the integrity of this system in AD.

While the mechanism(s) responsible for the degeneration of BFCNs is yet to be defined fully, there is evidence in AD to support the assertion that NGF signaling is implicated. As for rodents, the human hippocampus expresses the gene for NGF, human BFCNs express TrkA, the receptor tyrosine kinase for NGF, and these neurons respond to NGF in vitro and in vivo (Salehi et al. 2007a). In rodents, BFCNs are dependent on NGF for survival in early development and for maintenance in maturity (Sofroniew et al. 2001). Among the phenotypes that attend NGF deprivation in rodents is the atrophy of BFCN cell bodies. Furthermore, mouse models producing antibodies to NGF demonstrate a variety of neuropathological features of AD, including severe BFCN degeneration (Capsoni et al. 2000). In AD, NGF protein levels are increased in the BFCN projection sites, i.e., the hippocampal and cortical regions and, as evidenced through studies of immunostaining, are decreased in BFCN cell bodies. This finding suggests a defect in NGF retrograde transport. As might be expected, in view of the positive effect of NGF on the synthesis of its TrkA receptor (Holtzman et al. 1992), the levels of this protein are decreased in BFCNs in AD. Interestingly, in animal studies, NGF infusions reversed or limited the effects of severing the fimbria-fornix, i.e., BFCN axons projecting to the hippocampus. These and other studies have suggested the therapeutic potential for delivery of NGF to nucleus basalis. Preliminary data have indicated beneficiary effects (Tuszynski et al. 2005).

1.7 Using Mouse Models to Uncover the Molecular Mechanisms of Cholinergic Hippocampal De-afferentation

DS in the most common cause of mental retardation in children (Salehi et al. 2008; Roizen and Patterson 2003) and is caused by complete or partial triplication of chromosome 21. Trisomy 21 is the most common viable form of trisomy in humans. There are at least 364 known and predicted genes on HSA21 (Hattori et al. 2000). DS features include typical facial abnormalities, hypotonia, mental retardation, and cardiac abnormalities. Nervous system involvement, which affects patients throughout the lifespan, results in deficits involving learning, memory and language. Interestingly, after age 40, there is a striking similarity between AD and DS neuropathology (Wisniewski et al. 1985), and a majority of people with DS have further cognitive decline in their seventh decade (Chapman and Hesketh 2000). Thus, DS consistently activates pathogenetic mechanisms that lead to AD.

A majority of HSA21 orthologues have been mapped to the distal end of mouse chromosome 16 (MMU16). For this reason, a mouse has been developed that is segmentally trisomic for this portion of MMU16, the Ts65Dn mouse. Ts65Dn mice have three copies of a fragment of MMU16 extending from *Gabpa* to *Mx1* (Salehi et al. 2007b). In behavioral analyses, Ts65Dn mice reveal significant spatial learning disabilities, as shown by hidden platform and probe tests in the Morris water maze (Sago et al. 2000). Furthermore, Ts65Dn mice recapitulate a variety of DS morphological changes, including synaptic structural abnormalities in territories that receive BFCN projections (Belichenko et al. 2004, 2007).

Our investigations showed that failed axonal transport in Ts65Dn mice precedes BFCN degeneration. Young adult (6-month-old) Ts65Dn mice do show signs of atrophy or loss of marker. However, these mice show a significant reduction in the size and number of p^{75NTR}-labeled BFCNs at the age of 12 months. We found reduced NGF axonal transport in Ts65Dn mice as early as 3 months.

In a series of experiments, we studied the status of NGF gene expression and signaling in Ts65Dn and their 2N controls. We found a dramatic reduction in the retrograde transport of NGF in young adult Ts65Dn mice (Cooper et al. 2001). This reduction appeared to be somewhat selective since there was no decrease in the retrograde transport of fluorogold (Salehi et al. 2006), a molecule widely used to examine non-specific retrograde transport (Wessendorf et al. 1991). Ts1Cje mice are trisomic for a shorter segment of MMU16 that extends from *Sod1* to *Mx1* ($\sim$100 genes homologous to those on HSA21). NGF transport in Ts1Cje mice is significantly improved relative to that in the Ts65Dn mouse. Correspondingly, unlike Ts65Dn mice, NGF protein levels in Ts1Cje mice were similar to those of 2N mice in the hippocampus and septum (Salehi et al. 2006). Importantly, no significant changes could be found in the size or number of BFCNs in the MSN of these mice even in old age (Fig. 2). Recent data from Chen and colleagues (2008) have supported these findings. Using quantitative magnetic resonance imaging (MRI) in 2N, Ts65Dn and Ts1Cje mice, it was found that BFCN cell bodies in Ts65Dn, but not in

Ts1Cje mice, generated a significantly reduced signal [transverse proton spin-spin [T (2)] relaxation time].

These data prompted us to conclude that one or more genes in the segment that distinguishes Ts65Dn and Ts1Cje mice are necessary for the dramatic reduction of NGF transport. Due to the following, we chose to study the role of App overexpression in failed NGF transport.

1) a significant improvement in NGF transport in Ts1Cje mice monosomic for a segment of MMU16 with *App* (Salehi et al. 2006).
2) *APP* mutations lead to a familial form of AD (Goate et al. 1991).
3) *APP* duplication leads to a familial form of AD with major vascular pathology (Rovelet-Lecrux et al. 2006).
4) The need for the presence of the *APP*-containing region in HSA21 for development of AD-related pathology in an elderly woman with DS (Prasher et al. 1998).

Based on these findings, we chose to study the effects of *App* overexpression on axonal transport.

1.8 Role of APP in Failed NGF Axonal Transport

Comparing Ts65Dn mice trisomic (Ts65Dn: *App*+/+/+) with disomic (Ts65Dn: *App*+/+/−) mice for *App* revealed that Ts65Dn mice, with only two copies *App*, displayed a significant improvement in NGF transport. Thus, deleting one copy of *App* markedly improved NGF retrograde transport in Ts65Dn mice. These data were supported by the finding of significant negative correlation between NGF transport and hippocampal App-CTF levels. Thus, there is evidence that increased *App* gene dosage is necessary for the decrease in transport and degeneration of BFCNs (Fig. 2).

Our studies also provided evidence that NGF axonal transport was significantly diminished in APP_{Swe} mice and even more so doubly Tg mice. Furthermore, mice expressing entire human wild type *APP* (Lamb et al. 1993) showed a similar decline in transport. Thus, even a modest increase in the levels of *APP* leads to a significant decline in NGF retrograde transport. These data are evidence that an increased *App* gene dose is also sufficient for the decrease in NGF transport.

1.9 Early Endosomes and Their Role in Failed NGF Axonal Transport

EEs are intracellular organelles with a diameter of 50 nm that are involved in NGF retrograde transport (Delcroix et al. 2003). Moreover, increased EE size has been

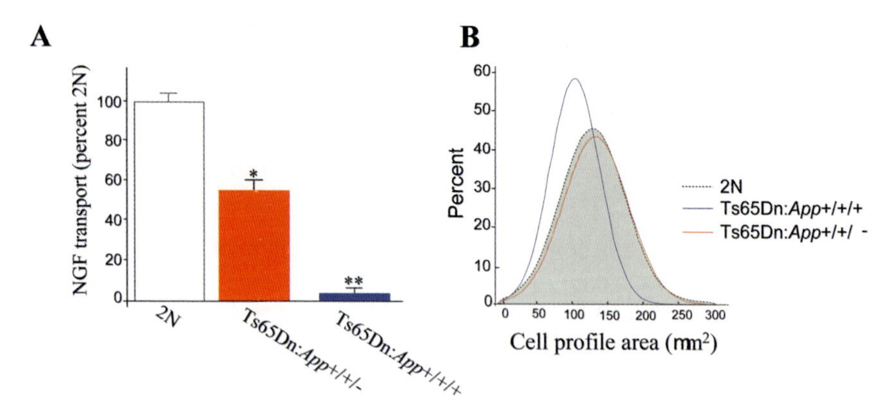

Fig. 2 (**A**) NGF transport in Ts65Dn: *App* $+/+/+$ mice as compared to controls (2N); $p < 0.0001$). There was a significant improvement in NGF axonal transport in Ts65Dn: *App* $+/+/-$; a highly significant change ($p = 0.0005$). (**B**) The BFCN atrophy in Ts65Dn: *App* $+/+/+$ mice was not present in Ts65Dn:*App* $+/+/-$ mice. Comparing the frequency distribution of BFCN cell profile areas, there was a significant difference ($p = 0.045$) between Ts65Dn: *App* $+/+/+$ and Ts65Dn: *App* $+/+/-$. (From Salehi et al. 2006, with permission from Elsevier)

reported in both DS and early AD (Cataldo et al. 2003; Cui et al. 2007). Accordingly, we reasoned that EEs might be important in the pathogenesis of failed axonal transport in Ts65Dn mice. Our previous studies indicated that NGF is found in EEs in cholinergic terminals in the hippocampus. Furthermore, these NGF-containing EEs are enlarged in BFCN terminals in the Ts65Dn hippocampus. At the present time, we are developing methods (see below) to study whether or not abnormal EEs are responsible for the defect in NGF transport and, if so, what role overexpression of App plays in causing this abnormality.

1.10 Methods to Study NGF Transport in Living Cells

To gain insight into the mechanisms by which NGF signaling endosomes are trafficked within axons, we have recently developed novel techniques to label NGF. Dorsal root ganglia (DRGs) have NGF signaling similar to that of BFCNs, are readily available for study and appear to be abnormal in people with DS. For these reason, we studied NGF transport in DRGs.

To study axonal transport, we made use of innovative tools: (1) a compartmented culture chamber (Fig. 3; Taylor et al. 2006) in which labeled NGF can be added to the distal axon chamber, and (2) the trafficking of NGF-containing endosomes tracked through the use of pseudo-total internal reflection fluorescence (pseudo-TIRF) microscopy. Trafficking of NGF is visualized through the conjugation of biotinylated NGF with Quantum dots (QD-NGF). Dissociated neurons are seeded in the cell body chamber. Axons generally grow through the microgrooves and reach the distal axon chamber. The culture system allows us to manipulate expression of

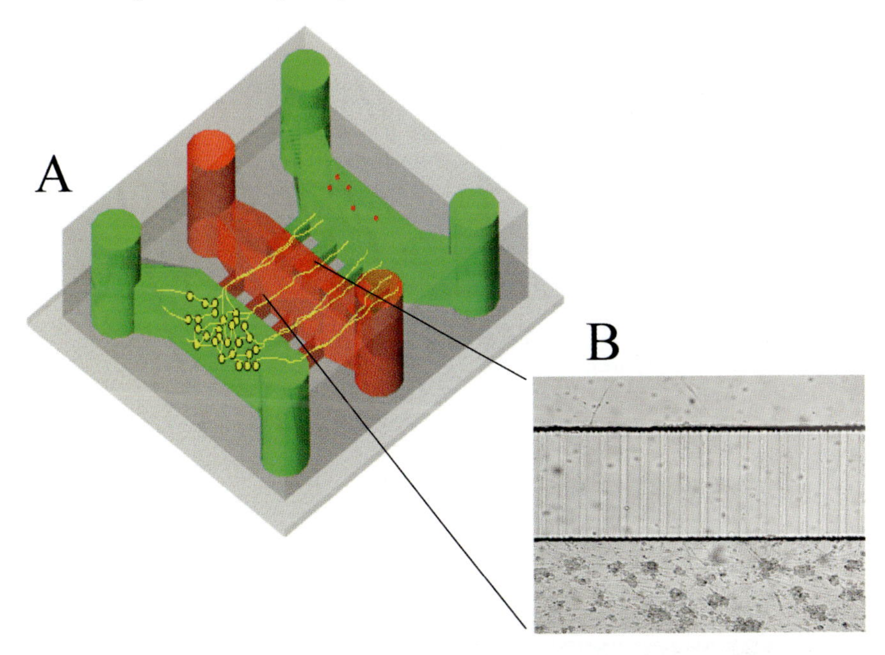

Fig. 3 (**A**) Schematic representation of a compartmented micro-fluid chamber. Cell body, axons and axon terminals are in different compartments. (**B**) A micrograph depicting DRGs in the compartmented micro-fluid chamber

genes of interest (e.g., *APP*) and to determine effects on the pattern, rate and amount of retrograde transport of NGF and NGF signaling.

QD605-NGF-containing endosomes often exhibit a pattern of movement that features movement followed by pauses. Almost all movement was in the retrograde direction. Examined across many examples, the movement of endosomes containing NGF resembled multi-lane highway traffic. Most endosomes moved independently of one another: fast moving ones passed those moving more slowly or those that had paused. We also noted examples in which paused endosomes appeared to obstruct the advance of other endosomes. Occasionally, two or more endosomes located very near one another travelled at the same speed for a few seconds before eventually separating (Fig. 4).

The number of endosomes observed in a fixed length of axon increased significantly with increased QD605-NGF concentration, ranging from 5 to 500 ng/ml (Fig 4), suggesting that the endosomal system has a capacity that exceeds that which would be occupied by NGF at concentrations in the physiological range. We detected no significant change in the stop-and-go pattern of movement, or the average speed of movement, of endosomes at increasing QD605-NGF concentrations. QD605-NGF-containing endosomes were readily detected at 5 ng/ml, a concentration that induced a robust neurite outgrowth response in pheochromocytoma cells (PC12) cells. The distance between adjacent QD605-NGF endosomes under this condition averaged about 69 μm. With increasing QD605-NGF concentration, the

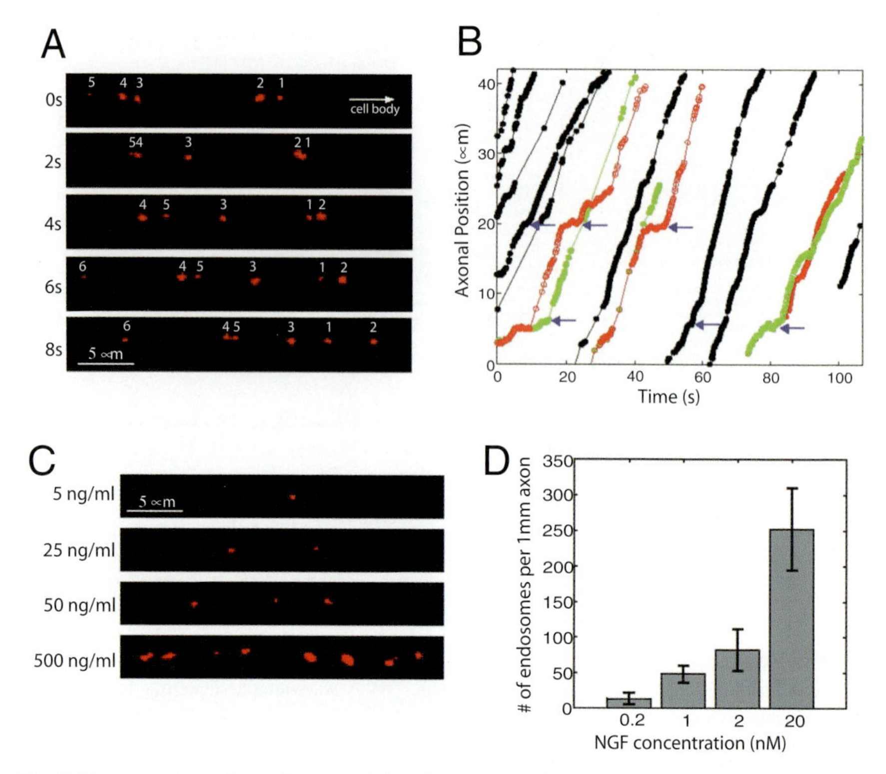

Fig. 4 Transport dynamics and concentration dependence of QD-NGF containing endosomes. (**A**) Time-lapse video images of endosomes traveling on the same axon. Five endosomes were visible at the beginning of the video recording, and the sixth endosome came into the field of view after 6 s. The white arrow indicates that direction of motion was toward the cell body. (**B**) Trajectories of 15 endosomes moving in the same axon through the same field of view. The majority of endosomes moved independently (black circles). Endosomes moving together or passing another endosome are shown in red and green for clarity. The blue arrows indicate the places where some trajectories paused at the same axonal location. (**C**) The number of endosomes in a fixed length of axon increases with QD-NGF concentration. (**D**) Average number of endosomes per 1 mm of axon increases with increased QD-NGF concentration ranging from 0.2 to 20 nM. (From Cui et al. 2007, with permission form PNAS)

number of endosomes traveling in the axon also increased (Fig. 4C). The number of endosomes per 1 mm of axon was estimated to be ~14 at QD605-NGF concentration at 5 ng/ml, ~49 at 25 ng/ml, ~83 at 50 ng/ml and ~252 at 500 ng/ml (Fig. 4D).

The photo-blinking property of QD605 fluorescence (Hohng and Ha 2004; i.e., on-off-on fluorescence emission) allowed us to determine the number of QD605-NGF molecules per endosome. At our experimental conditions (532 nm green laser excitation), QD605 spent about 5–10% of time in a dark state that did not emit fluorescent light. Endosomes containing a single QD605 were identified individually by checking for the blinking events longer than 5 consecutive frames (0.5 s) during their movement. For endosomes that did not blink, the number of QD605-NGF complexes present was determined by comparing the fluorescence intensity to that for endosomes containing a single QD605-NGF that did blink. However, this number is an approximation, due to the variation in the fluorescence intensity of a single QD605 ($\sim$70%). Using these measures, the majority of endosomes seen when cultures were treated with an effective concentration of NGF of 1 nM contained a single QD605, of which 95% exhibited characteristic photo blinking. Importantly, immunostaning of fixed cultures showed that NGF was transported, together with its TrkA receptors, in EEs marked by Rab5, and with the activated form of Erk1/2. These results point to the ability to reliably label moving EEs, to explore the dynamics of their movement, to define their NGF content, and to interrogate their signaling and delivery to the cell body. The ability to make such measurements will greatly facilitate studies to define the mechanism(s) by which APP compromises endosomal transport and to answer the questions posed above regarding its significance for the degeneration of BFCNs and, perhaps, other neurons whose age-related degeneration characterizes DS and AD.

1.11 Conclusions

Precise anatomical analyses of affected systems have revealed that all send extensive projections to the hippocampus. A large number of experiments in animal models have shown that lesions to the tracts linking subcortical systems to the hippocampus lead to de-afferentation of the hippocampus, with serious anatomical, electrophysiological and behavioral consequences. During the last decade, our studies on AD and DS have shown that failed trafficking of NGF signals represents an attractive hypothesis to explain degeneration of BFCNs and perhaps other affected neurons and the hippocampus, and these findings propose new therapies.

Acknowledgements This study was supported by grants from the Alzheimer Association, State of California Alzheimer's Program, and Thrasher Foundation.

References

Belichenko PV, Masliah E, Kleschevnikov AM, Villar AJ, Epstein CJ, Salehi A, Mobley WC (2004) Synaptic structural abnormalities in the Ts65Dn mouse model of Down syndrome. J Comp Neurol 480:281–298

Belichenko PV, Kleschevnikov AM, Salehi A, Garner C, Mobley W (2007) Synaptic and cognitive abnormalities in mouse models of Down syndrome: exploring genotype-phenotype relationships. J Comp Neurol 504:329–345

Borelli KG, Gargaro AC, dos Santos JM, Brandao ML (2005) Effects of inactivation of serotonergic neurons of the median raphe nucleus on learning and performance of contextual fear conditioning. Neurosci Lett 387:105–110

Braak E, Griffing K, Arai K, Bohl J, Bratzke H, Braak H (1999) Neuropathology of Alzheimer's disease: what is new since A. Alzheimer? Eur Arch\Psych Clin Neurosci 249 S3:14–22

Capsoni S, Ugolini G, Comparini A, Ruberti F, Berardi N, Cattaneo A (2000) Alzheimer-like neurodegeneration in aged antinerve growth factor transgenic mice. Proc Natl Acad Sci USA 97:6826–6831

Cataldo AM, Petanceska S, Peterhoff CM, Terio NB, Epstein CJ, Villar A, Carlson EJ, Staufenbiel M, Nixon RA (2003) App gene dosage modulates endosomal abnormalities of Alzheimer's disease in a segmental trisomy 16 mouse model of Down syndrome. J Neurosci 23:6788–6792

Celada P, Siuciak JA, Tran TM, Altar CA, Tepper JM (1996) Local infusion of brain-derived neurotrophic factor modifies the firing pattern of dorsal raphe serotonergic neurons. Brain Res 712:293–298

Chapman RS, Hesketh LJ (2000) Behavioral phenotype of individuals with Down syndrome. Mental Retardation Dev Disabilities Res Rev 6:84–95

Chen Y, Dyakin VV, Branch CA, Ardekani B, Yang D, Guilfoyle DN, Peterson J, Peterhoff C, Ginsberg SD, Cataldo AM, Nixon RA (2008) In vivo MRI identifies cholinergic circuitry deficits in a Down syndrome model. Neurobiol Aging doi:10.1016

Compton DM, Dietrich KL, Smith JS, Davis BK (1995) Spatial and non-spatial learning in the rat following lesions to the nucleus locus coeruleus. Neuroreport 7:177–182

Cooper JD, Salehi A, Delcroix JD, Howe CL, Belichenko PV, Chua-Couzens J, Kilbridge JF, Carlson EJ, Epstein CJ, Mobley WC (2001) Failed retrograde transport of NGF in a mouse model of Down's syndrome: reversal of cholinergic neurodegenerative phenotypes following NGF infusion. Proc Natl Acad Sci USA 98:10439–10444

Cui B, Wu C, Chen L, Ramirez A, Bearer EL, Li WP, Mobley WC, Chu S (2007) One at a time, live tracking of NGF axonal transport using quantum dots. Proc Natl Acad Sci USA 104:13666–13671

Delcroix JD, Valletta JS, Wu C, Hunt SJ, Kowal AS, Mobley WC (2003) NGF signaling in sensory neurons: evidence that early endosomes carry NGF retrograde signals. Neuron 39:69–84

Delcroix JD, Valletta J, Wu C, Howe CL, Lai CF, Cooper JD, Belichenko PV, Salehi A, Mobley WC (2004) Trafficking the NGF signal: implications for normal and degenerating neurons. Prog Brain Res 146:3–23

Ginty DD, Segal RA (2002) Retrograde neurotrophin signaling: Trk-ing along the axon. Curr Opin Neurobiol 12:268–274

Goate A, Chartier-Harlin MC, Mullan M, Brown J, Crawford F, Fidani L, Giuffra L, Haynes A, Irving N, James L et al. (1991) Segregation of a missense mutation in the amyloid precursor protein gene with familial Alzheimer's disease. Nature 349:704–706

Hattori M, Fujiyama A, Taylor TD, Watanabe H, Yada T, Park HS, Toyoda A, Ishii K, Totoki Y, Choi DK, Groner Y, Soeda E, Ohki M, Takagi T, Sakaki Y, Taudien S, Blechschmidt K, Polley A, Menzel U, Delabar J, Kumpf K, Lehmann R, Patterson D, Reichwald K, Rump A, Schillhabel M, Schudy A, Zimmermann W, Rosenthal A, Kudoh J, Schibuya K, Kawasaki K, Asakawa S, Shintani A, Sasaki T, Nagamine K, Mitsuyama S, Antonarakis SE, Minoshima S, Shimizu N, Nordsiek G, Hornischer K, Brant P, Scharfe M, Schon O, Desario A, Reichelt J, Kauer G, Blocker H, Ramser J, Beck A, Klages S, Hennig S, Riesselmann L, Dagand E, Haaf T, Wehrmeyer S, Borzym K, Gardiner K, Nizetic D, Francis F, Lehrach H, Reinhardt R, Yaspo ML (2000) The DNA sequence of human chromosome 21. Nature 405:311–319

Heerssen HM, Segal RA (2002) Location, location, location: a spatial view of neurotrophin signal transduction. Trends Neurosci 25:160–165

Hohng S, Ha T (2004) Near-complete suppression of quantum dot blinking in ambient conditions. J Am ChemSoc 126:1324–1325

Holtzman DM, Li Y, Parada LF, Kinsman S, Chen CK, Valletta JS, Zhou J, Long JB, Mobley WC (1992) p140trk mRNA marks NGF-responsive forebrain neurons: evidence that trk gene expression is induced by NGF. Neuron 9:465–478

Howe CL, Mobley WC (2005) Long-distance retrograde neurotrophic signaling. Curr Opin Neurobiol 15:40–48

Moreau PH, Cosquer B, Jeltsch H, Cassel JC, Mathis C (2008) Neuroanatomical and behavioral effects of a novel version of the cholinergic immunotoxin mu p75-saporin in mice. Hippocampus 18:610–622

Morrison JH, Hof PR (2002) Selective vulnerability of corticocortical and hippocampal circuits in aging and Alzheimer's disease. Prog Brain Res 136:467–486

Mufson EJ, Kroin JS, Sendera TJ, Sobreviela T (1999) Distribution and retrograde transport of trophic factors in the central nervous system: functional implications for the treatment of neurodegenerative diseases. Prog Neurobiol 57:451–484

Prasher VP, Farrer MJ, Kessling AM, Fisher EM, West RJ, Barber PCButler AC (1998) Molecular mapping of Alzheimer-type dementia in Down's syndrome. Ann Neurol 43:380–383

Rogaeva E, Meng Y, Lee JH, Gu Y, Kawarai T, Zou F, Katayama T, Baldwin CT, Cheng R, Hasegawa H, Chen F, Shibata N, Lunetta KL, Pardossi-Piquard R, Bohm C, Wakutani Y, Cupples LA, Cuenco KT, Green RC, Pinessi L, Rainero I, Sorbi S, Bruni A, Duara R, Friedland RP, Inzelberg R, Hampe W, Bujo H, Song YQ, Andersen OM, Willnow TE, Graff-Radford N, Petersen RC, Dickson D, Der SD, Fraser PE, Schmitt-Ulms G, Younkin S, Mayeux R, Farrer LA, St George-Hyslop P (2007) The neuronal sortilin-related receptor SORL1 is genetically associated with Alzheimer disease. Nature Genet 39:168–177

Roizen NJ, Patterson D (2003) Down's syndrome. Lancet 361:1281–1289

Rovelet-Lecrux A, Hannequin D, Raux G, Le Meur N, Laquerriere A, Vital A, Dumanchin C, Feuillette S, Brice A, Vercelletto M, Dubas F, Frebourg TCampion D (2006) APP locus duplication causes autosomal dominant early-onset Alzheimer disease with cerebral amyloid angiopathy. Nature Genet 38:24–26

Sago H, Carlson EJ, Smith DJ, Rubin EM, Crnic LS, Huang TT, Epstein CJ (2000) Genetic dissection of region associated with behavioral abnormalities in mouse models for Down syndrome. Ped Res 48:606–613

Salehi A, Pohlman B, Mobley WC (2008) Down syndrome/trisomy 21. Encycl Neurosci, in press.

Salehi A, Delcroix JD, Belichenko PV, Zhan K, Wu C, Valletta JS, Takimoto-Kimura R, Kleschevnikov AM, Sambamurti K, Chung PP, Xia W, Villar A, Campbell WA, Kulnane LS, Nixon RA, Lamb BT, Epstein CJ, Stokin GB, Goldstein LS, Mobley WC (2006) Increased App expression in a mouse model of Down's syndrome disrupts NGF transport and causes cholinergic neuron degeneration. Neuron 51:29–42

Salehi A, Kleschevnikov AM, Mobley W (2007a) Cholinergic neurodegeneration in Alzheimer's disease: basis for nerve growth factor therapy. In: Cuello AC.(ed) Pharmacological mechanisms in Alzheimer's therapeutics. Springer 64–104

Salehi A, Faizi M, Belichenko PV, Mobley WC (2007b) Using mouse models to explore genotype-phenotype relationship in Down syndrome. Mental Retardation De Disabilities Res Rev 13:207–214

Schmidt V, Sporbert A, Rohe M, Reimer T, Rehm A, Andersen OMWillnow TE (2007) SorLA/LR11 regulates processing of amyloid precursor protein via interaction with adaptors GGA and PACS-1. J Biol Chem 282:32956–32964

Shahidi S, Motamedi F, Naghdi N (2004) Effect of reversible inactivation of the supramammillary nucleus on spatial learning and memory in rats. Brain Res1026:267–274

Sofroniew MV, Howe CL, Mobley WC (2001) Nerve growth factor signaling, neuroprotection, and neural repair. Annu Rev Neurosci 24:1217–1281

Taylor AM, Rhee SW, Jeon NL (2006) Microfluidic chambers for cell migration and neuroscience research. Methods Mol Biol 321:167–77

Tuszynski MH, Thal L, Pay M, Salmon DP, U HS, Bakay R, Patel P, Blesch A, Vahlsing HL, Ho G, Tong G, Potkin SG, Fallon J, Hansen L, Mufson EJ, Kordower JH, Gall C, Conner J (2005) A phase 1 clinical trial of nerve growth factor gene therapy for Alzheimer disease. Nature Med 11:551–555

Wessendorf MW (1991) Fluoro-Gold: composition, and mechanism of uptake. Brain Res 553:135–148

Wisniewski KE, Dalton AJ, McLachlan C, Wen GY, Wisniewski HM (1985) Alzheimer's disease in Down's syndrome: clinicopathologic studies. Neurology 35:957–961

Wu C, Ramirez A, Cui B, Ding J, Delcroix JD, Valletta JS, Liu JJ, Yang Y, Chu S, Mobley WC (2007) A functional dynein-microtubule network is required for NGF signaling through the Rap1/MAPK pathway. Traffic 8:1503–1520

Membrane Trafficking and Targeting in Alzheimer's Disease

Lawrence Rajendran(✉) and Kai Simons

Abstract The key players in the processing of the amyloid precursor protein (APP), i.e., α-, β-, γ-secretase and the substrate APP, are all membrane associated and hence are subjected to regulation by the lipid environment and membrane trafficking. This review focuses on how membrane-associated events regulate amyloidogenic processing of APP and discusses ways to design membrane trafficking-based strategies to interfere with the process.

1 Introduction

Two distinguishing features of Alzheimer's disease (AD) are the presence of neurofibrillary tangles and amyloid plaques. Amyloid plaques contain the β-amyloid peptide (Aβ), which in either its plaque-associated form or a soluble oligomeric form is thought to set in a cascade of events that eventually lead to neurodegeneration. Aβ is derived from a large type I transmembrane protein, the amyloid precursor protein (APP; Selkoe et al. 1996). APP is cleaved sequentially by enzymes termed β- and γ-secretase. β-Secretase activity is conferred by the enzyme, β-APP cleaving enzyme (BACE-1, hereafter referred to β-secretase), which cleaves APP in its luminal domain to generate a secreted ectodomain (sAPPβ) and a C-terminal fragment of APP (β-CTF). The latter fragment subsequently becomes a substrate for the membrane-bound enzymatic complex termed γ-secretase, which cleaves the transmembrane domain of β-CTF to release the lumenal Aβ and a cytoplasmic soluble fragment, termed AICD (APP intracellular domain; Annaert and De Strooper 2002; Selkoe et al. 1996; Small and Gandy 2006). γ-Secretase is a multiprotein complex consisting of the catalytic presenilins-1 and -2 for activity (Annaert and

L. Rajendran and K. Simons
Max Planck Institute of Molecular Cell Biology and Genetics, Pfotenhauerstr. 108, 01307, Dresden, Germany

P. St. George-Hyslop et al. (eds.) *Intracellular Traffic and Neurodegenerative Disorders*, Research and Perspectives in Alzheimer's Disease,

De Strooper 2002) along with accessory proteins such as nicastrin, Aph-1, and PEN-2 (De Strooper 2003; Edbauer et al. 2003). APP can also be alternatively cleaved by a non-amyloidogenic, transmembrane enzyme called α-secretase that cleaves APP inside the Aβ region, thus precluding the formation of Aβ (Kojro and Fahrenhol 2005). Thus, the core proteins involved in APP processing, i.e., APP, α-, β- and γ-secretase, are all membrane associated and hence are subjected to regulation by the lipid environment and membrane trafficking processes (Hooper 2005).

2 Role of Lipids in the Amyloidogenic Cleavage of APP

The enzymatic activity of membrane-associated enzymes is regulated mainly by the pH of the lumenal aqueous environment, ionic strength and the nature of the lipids that are found in its immediate neighborhood in the membrane plane. Membrane lipids could serve either as cofactors or co-structures or could provide optimal bulk membrane properties that in turn modulate the activity of the enzyme. β-Secretase is no exception to this modulation, and the modulation is mainly mediated by a class of lipids called raft lipids, i.e., cholesterol and sphingolipids. Rafts are lateral assemblies of sphingolipids and cholesterol within the membrane (Rajendran and Simons 2005). These lipids tend to assemble laterally in the membrane, thereby forming ordered regions that segregate from the liquid-disordered matrix of the cellular membrane. A fraction of β-secretase is associated with lipid rafts in a cholesterol-dependent manner (Cordy et al. 2003; Ehehalt et al. 2003; Simons et al. 2001), and it is in these domains that β-secretase is thought to cleave APP. Cholesterol depletion, either by statins or cyclodextrin, reduced β-secretase activity (Ehehalt et al. 2003). However, upon inhibition of β-secretase activity by cholesterol depletion, the activity of the non-raft-associated α-secretase was elevated, suggesting that, upon disruption of raft domains, the fraction of non-raft APP was increased and became more available for cleavage by α-secretase (Kojro et al. 2001). Studies with purified β-secretase reconstituted into proteoliposomes confirmed that cerebrosides and cholesterol activate the enzyme (Kalvodova et al. 2005). These results led to the interpretation that APP is present in two pools in the membrane, one associated with lipid rafts, where β- and γ-cleavages of APP occur, and another outside of rafts, where α-cleavage occurs. This model of lateral segregation offers an explanation as to how the same protein, in this case, APP, can be processed in two different mutually exclusive ways (Simons et al. 2001).

What about γ-secretase? γ-Secretase is also shown to partition into raft domains, and its activity towards cleaving β-CTF is cholesterol dependent (Lee et al. 1998; Vetrivel et al. 2004). By reconstituting γ-secretase complex and β-CTF in various liposomes differing in lipid composition, Dennis Selkoe's lab recently showed that γ-secretase functioned best in a lipid environment that contained cholesterol and sphingolipids, similar to our findings on β-secretase (Kalvodova et al. 2005; Osenkowski et al. JBC 2008, in press). Hence one could envision that the entire amyloidogenic pathway is somehow dependent on these cholesterol-enriched raft domains. Whether there are two pools of γ-secretase, one associated with rafts that cleave β-CTF and another non-raft associated that cleaves α-CTF, is still unclear. An equally interesting aspect to study is the raft dependence of γ-secretase towards other substrates, Notch.

3 Site of Amyloidogenic Cleavages of APP in the Endocytic Pathway

While lateral segregation is important in determining the accessibility of APP to secretases, subcellular compartmentalization of the secretases also this phenomenon. β-Secretase, owing to its low pH requirement, needs an acidic compartment for its enzymatic activity whereas α-secretase is shown to be active at the plasma membrane (Kalvodova et al. 2005). Hence APP at the plasma membrane could be cleaved by α-secretase whereas the pool that reaches the intracellular acidic compartment could be cleaved by β-secretase, and several reports support this hypothesis (Daugherty and Green 2001; Ehehalt et al. 2003; Kinoshita et al. 2003; Kojro et al. 2001; Refolo et al. 1995).

Site of β-cleavage: To identify the subcellular compartment in the endocytic pathway where β-cleavage occurs, we used specific antibodies against the β-cleaved ectodomain. By colocalizing the β-cleaved ectodomain along with endogenous endocytic markers, we found that the β-cleaved ectodomain mainly co-localized with an early endosomal marker, EEA-1. We also used GFP fusions of small GTPase proteins called rab proteins as markers for the endosomes (rab5 for the early endosomes, rab7 as a late endosome marker and rab11 for the recycling compartment). Colocalization studies suggested that early and late endosomes were the earliest stations where β-cleaved products accumulated. To understand if cleavage could happen in early endosomes, we overexpressed the GTPase mutant of rab5 (rab5Q79L), which inhibited cargo (in this case, APP and β-secretase) flow from early to late endosomes (Rink et al. 2005) to confine the cargo in the early endosomal compartment. Under these conditions, we observed that almost all of the cellular β-cleaved ectodomain was sequestered in these enlarged early endosomes, consistent with the idea that the early endosomal sorting of APP/β-secretase is sufficient for β-cleavage to occur. All these experiments were performed under steady-state conditions. To specifically follow the endocytosis of plasma membrane-associated APP and β-secretase, we performed induced endocytosis experiments, where APP and β-secretase were labeled with antibodies at the cell surface and internalized. After 5 minutes of internalization, APP and β-secretase co-localized in rab5-positive early endosomes. Fluorescence resonance energy transfer (FRET) measurements also confirmed that maximum intermolecular interaction between APP and β-secretase occurred in the early endosomes (our unpublished results and Kinoshita et al. 2003). These data and results from live imaging of surface-labeled

APP and β-secretase suggested that early endosomes were a major site of β-cleavage (Rajendran et al. 2006).

Site of γ-cleavage: Aβ production from APP requires the additional enzymatic cleavage by γ-secretase complex. To identify the compartment where γ-cleavage of APP occurs, we used the β-CTF (C99 fragment) fused with GFP as a substrate for γ-secretase. This GFP fragment, under normal conditions, showed localization mainly to the Golgi complex and a diffused staining pattern all over the cell, due to the fact that β-CTF was produced in the biosynthetic pathway, matured through the Golgi and reached a post-Golgi compartment where it was cleaved by γ-secretase to release the soluble fragment, AICD. Since GFP was fused to the C-terminal of β-CTF, AICD harbored the GFP and hence a diffuse GFP signal was seen that permeated the whole cell (including the nucleus). However, once γ-secretase activity was inhibited by DAPT, a γ-specific inhibitor, a dramatic difference in the β-CTF staining was observed. β-CTF now showed a punctuated pattern, suggesting that, under γ-secretase-inhibited conditions, the substrate accumulated in a particular organelle where γ-cleavage would have otherwise occurred. This reasoning led us to the identification of early endosomes as sites for γ-cleavage of APP. Our data confirmed data from Christoph Kaether and Christian Haass, in Munich, who also demonstrated that γ-cleavage of β-CTF occurred in early endosomes (Kaether et al. 2006).

Together, these results showed that early endosomes were a major cellular site for Aβ production (Fig. 1). Hence mechanisms that control the cargo flow to and away from the early endosomes should regulate APP processing. Interestingly, we found that stimulation of recycling of cargo from the early endosome back to the plasma membrane by overexpression of wild type rab4 (de Renzis et al. 2002) decreased β-cleavage and Aβ secretion, suggesting that reducing the residence time of APP and/or β-secretase in early endosomes reduced β-cleavage (Rajendran et al. 2006). Proteins that belong to the retromer family retrieve cargo from early endosomes to Golgi and have been recently implicated in AD (Small and Gandy 2006). Mutations in SorLA lead to the failure of retrieving APP from early endosomes and consequentially increase Aβ production (Andersen et al. 2005; Rogaeva et al. 2007). On the other hand, proteins of the GGA family have been shown to transport β-secretase from endosomes to Golgi, and depletion of GGA proteins leads to increased amyloidogenic processing of APP (He et al. 2005; Tesco et al. 2007). Evidently membrane trafficking to and away from early endosomes regulates the residency of the cargoes (i.e., of APP, β-secretase and the components of γ-secretase complex), and regulators of this process play an important role in amyloidogenic processing. In fact, a recent work from John Cirrito and David Holtzman group (Cirrito et al. 2008) showed that around 70% of Aβ is produced by the endocytic route in vivo. Identification of all components that regulate trafficking of APP and the secretase would greatly aid our understanding of the role of all the membrane-trafficking events in APP processing and bear relevance for AD diagnosis and therapy.

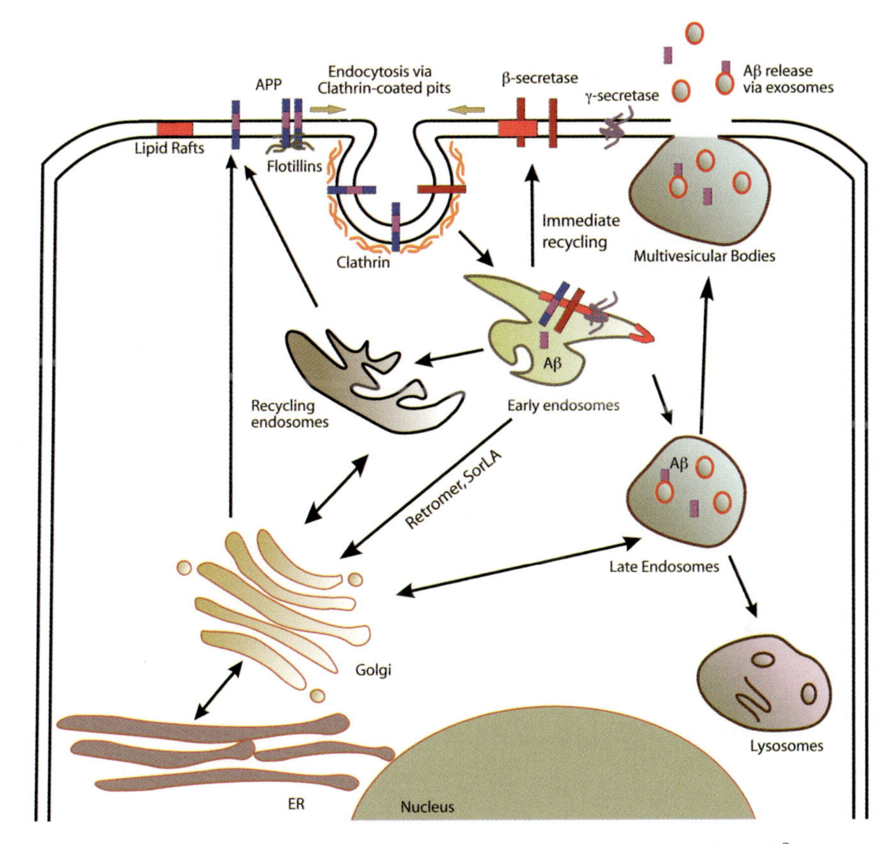

Fig. 1 Membrane-trafficking events in the amyloidogenic processing of APP. APP and β-secretase are internalized in a specialized, clathrin-mediated endocytic pathway that requires preclustering of APP via cholesterol and flotillins (Schneider et al. 2008). Internalized APP and β-secretase are sorted to early endosomes, where β-cleavage of APP occurs, to produce the soluble β-cleaved ectodomain and the c-terminal fragment, β-CTF. The latter becomes a substrate for γ-cleavage to produce Aβ and AICD. Aβ is then sorted to the multivesicular bodies, which fuse with the plasma membrane to release the intraluminal vesicles as exosomes. Some exosomes contain Aβ. Several trafficking routes going away from early endosomes are indicated

4 The Release of Aβ via Exosomal Pathway

Having shown that Aβ production occurred predominantly in early endosomes, we then addressed the question of Aβ secretion from these intracellular compartments. How does such an intracellularly generated, fairly hydrophobic Aβ released outside the cell contribute to the extracellular pool responsible for plaque formation? Some basic cell biology offered key insights to this process.

Proteins that are destined for lysosomal degradation are sorted from the early endosomes to the intraluminal vesicles (ILVs) of the multivesicular bodies (MVBs) or late endosomes. These multivesicular endosomes can then either fuse with lysosomes and degrade the cargo or fuse with the plasma membrane to release the ILVs as exosomes (Pelchen-Matthews et al. 2004). Since Aβ is generated in early endosomes in the endocytic route and is found in MVBs (Rajendran et al. 2006; Takahashi et al. 2002), we speculated that fusion of these MVBs with the plasma membrane could release Aβ complexed with exosomes (Raposo et al. 1996). By isolating exosomes from cell culture supernatants of cells that stably expressed either the wildtype or Swedish mutant of APP, we could show that indeed a small fraction of extracellular Aβ was associated with exosomes. Co-immuno-gold labeling with raft markers such as flotillin-1 and the ganglioside, GM1, showed that Aβ not only co-fractionated with exosomes but actually was also released in association with these vesicles (Rajendran et al. 2006). Since we observed only a small fraction being released via exosomes, we resorted to investigating two questions: (1) is there significance to the small fraction of Aβ that is released in association with these vesicles, and (2) how is the bulk of Aβ released?

When we performed immunohistochemistry analysis with antibodies against exosomal proteins on brain sections from AD patients, we observed an enrichment of exosomal proteins in the plaques. This observation was in agreement with our findings that Aβ peptides can be released complexed with exosomes from multivesicular bodies and led us to hypothesize that exosome-associated Aβ could be involved in plaque formation. Whether and how exosome-associated Aβ is causally linked to the pathogenesis of AD remains to be studied. But one can make some predictions based on data found in the literature. Since exosomes are enriched in glycolipids, raft lipids and raft-associated proteins such as GPI-anchored proteins (Fevrier et al. 2004; Thery et al. 2001), raft-bound Aβ could be incorporated into exosomes. The ganglioside, GM1-associated Aβ, has been shown to act as an amyloid seed for Aβ fibrillation (Hayashi et al. 2004), and the presence of GM1 and raft lipids (de Gassart et al. 2004) in exosomes bolsters the idea that exosomes can act as nucleation centers for plaque formation (Rajendran et al. 2006). In fact, work from Yanagisawa's lab recently showed that incubating Aβ with exosomes that contained the ganglioside, GM1, induced Aβ aggregation (Yuyama et al. 2008), thereby providing evidence that exosome-associated Aβ could act as a seed for plaque formation. The presence of certain lipids or proteins in the exosomes might aid in the oligomerization of Aβ. The case of a natural amyloid protein, Pmel17, is relevant for AD and worth mentioning here. This natural amyloid, found in melanosomes, is comprised of Pmel17, which is proteolytically processed in endosomes and the processed ectodomain (Mα) undergoes fibril formation in the lumen of MVBs. These MVBs are similar to those responsible for carrying exosomes (ESCRT independent; Raposo and Marks, 2007) and differ from those that carry cargoes for lysosomal degradation. In fact, it was recently shown that exosomal MVBs are enriched in ceramide, a lipid that also has its own relevance to AD, and are ESCRT independent (Trajkovic et al. 2008). Since Aβ is sorted to exosomal MVBs, we believe that these MVBs are similar to those in which Pmel17 undergoes amyloid formation, and the luminal vesicles (exosomes) could provide the right environment for Aβ to undergo fibril formation.

To address the issue of how the bulk of Aβ is released, we collaborated with Uwe Konietzko and Roger Nitsch, in Zurich. In early 2006, the Zurich team developed transgenic mice that expressed both the Swedish and Artic mutations in a single APP gene. They showed that these mice produced robust amounts of Aβ but did not exhibit any extracellular plaques; nevertheless, they exhibited cognitive disturbances (Knobloch et al. 2007). This finding gave us a handle to question where the bulk of Aβ was sequestered in the absence of the extracellular plaques. What we found was that much of the Aβ was localized to multivesicular bodies that were positive for the lipid raft-associated protein, flotillin-1 (Rajendran et al. 2007). On the one hand, these results showed that, preceding the detection of extracellular plaque formation, Aβ accumulated intracellularly and was sufficient to cause cognitive disturbances. On the other hand, they suggested that the bulk of Aβ that is produced in early endosomes was sorted to the multivesicular bodies to be released via the exosomal pathway. This pathway released a fraction of Aβ in association with exosomal vesicles and the remainder as soluble pool (Fig. 1). What causes Aβ to be released from the membrane plane is still not understood. After γ-cleavage, a negative charge is introduced at the C-terminus of Aβ that lies buried in the membrane. This charge introduction could make Aβ energetically unstable in the hydrophobic environment of the membrane and thus release it out of the membrane. Whether this release happens after the γ-cleavage in the early endosomes or outside the cell (due to pH differences, differences in the ionic strength in the extracellular milieu, etc.) is not yet understood. On the other hand, the soluble Aβ that is released from the cells could also hitch on exosomes in the extracellular milieu to mediate its intercellular transfer via exosomes.

5 Membrane Targeting for AD Therapy

Since β-cleavage of APP occurs in early endosomes, we reasoned that targeting active β-secretase in these endosomes would be a more efficient approach to inhibit the enzyme. As explained before, the pH of endosomes (pH 4.0–5.0) is optimal for β-secretase activity, which probably explains the requirement for endocytosis. Moreover, in view of the multiple functions of γ-secretase, β-secretase is still the preferred target for therapy (Vassar 2002). Several transition-state inhibitors have been designed to inhibit the active site of the enzyme (Hong et al. 2000; Tung et al. 2002), and since transition-state inhibitors recognize, bind to and inactivate only the active conformation of the enzyme, we speculated that, in order for these inhibitors to inhibit the active β-secretase found in the endosomes, we have to target the inhibition to these intracellular compartments. Hence we tested this hypothesis by examining the efficiency of a membrane-tethered version of an otherwise soluble inhibitor that is now targeted to endosomes via endocytosis (Rajendran et al. 2008). Membrane anchoring was achieved by linking the inhibitor to a cholesterol-like moiety via a linker. The linker length was chosen to be around 90Å, based on the β-cleavage site of APP. The results were clear: the free or anchorless inhibitor (the one that still contained the linker molecule but not the sterol) inhibited β-secretase in a cell-free enzymatic assay but did not inhibit β-cleavage in cells. On the other hand, β-secretase was effectively inhibited by the membrane-bound inhibitor, both in vitro and in vivo (transgenic flies and mice). Since β-secretase is also present on the plasma membrane but is active only when it gets internalized to the endosomes, the free inhibitor could not recognize the plasma membrane pool of the enzyme. Membrane anchoring, on the other hand, targeted the very same inhibitor to the endosomes. To demonstrate that the membrane-anchored inhibitor indeed gained access to the endosomes, we fluorescently labeled the free, anchorless inhibitor and the membrane-anchored inhibitor. After incubation of cells with the fluorescent inhibitors, we found that the membrane-anchored inhibitor readily partitioned into the membrane and subsequently internalized to endosome-like structures, whereas no internalization or membrane partitioning was observed in the free inhibitor. Subcellular localization revealed that this compartment is in the early endosomes and co-localization with internalized APP and β-secretase showed that the membrane-anchored inhibitor internalized to those endosomes where APP/β-secretase localized. This finding clearly showed that the transition-state inhibitor did not recognize the plasma membrane-associated β-secretase, and membrane anchoring rendered it endocytosis competent. These data confirmed the hypothesis that efficient inhibition of β-secretase requires targeting the inhibitors to the endosomes that house the active form of the enzyme (Rajendran et al. 2008).

The reasons why we used cholesterol-like anchor to promote inhibitory activity were two-fold: (1) membrane anchoring via sterol readily partitioned the inhibitor into the membrane plane, thereby trafficking the inhibitor to the endosomes, and (2) since β-secretase is enriched in cholesterol-raft domains, we speculated that linking the inhibitor to cholesterol would enrich the inhibitor in these domains. In this way, we wanted to address the issue of domain specificity. To determine whether targeting to raft domains promoted the inhibitory effect or if it was sufficient to localize the inhibitor in the membrane, we synthesized the inhibitors with different acyl anchors - palmitoyl, myristoyl and oleyl - that have different affinities for membranes and raft domains (Resh 2004). We measured the raft affinities of different anchored inhibitors using scanning fluorescence correlation spectroscopy and avalanche photodiode imaging on supported lipid bilayers (a model membrane system to study phases) that showed phase separation. We found that sterol-anchored inhibitors partitioned most to the raft-like structures, followed by palmitoyl- and oleyl-linked inhibitors. In agreement with our working hypothesis, sterol-linked inhibitors inhibited β-secretase most efficiently, followed by palmitoyl-, myristoyl- and then oleyl-linked inhibitors, suggesting that the higher the raft affinity, the more effective the inhibitors were. These results demonstrated that, by directing β-secretase inhibitors to raft microdomains within the endosomes, their inhibitory potential was enhanced (Rajendran et al. 2008).

Membrane anchoring thus dramatically increased the potency of a β-secretase inhibitor; we believe that this is primarily due to two reasons: (a) the inhibitor became endocytosis competent upon membrane anchoring and gained access to the endosomal β-secretase, and (b) we reduced the dimensionality of the otherwise soluble inhibitor. The free inhibitor is a soluble molecule that permeates the 3D space whereas membrane anchoring reduces its dimensionality to the 2D membrane plane, thereby increasing the effective concentration of the inhibitor in the target membrane. The reduction in the dimensionality of the inhibitor also enhanced the interaction between the inhibitor and the enzyme (Adam 1968). By confining the membrane anchor in the raft domains due to the sterol anchoring, we further enriched the inhibitor in the vicinity of raft-associated β-secretase, thus enhancing the interaction. The enhanced potency of the sterol-linked inhibitor supported our previous work, where we showed that both lipid environment and the subcellular localization of β-secretase regulated its activity (Ehehalt et al. 2003; Kalvodova et al. 2005; Rajendran et al. 2006).

We believe that these data open up the possibility that this approach could be used to design more effective β–secretase inhibitors for the treatment of AD. Whether these rational-based membrane-anchored inhibitors can pass the blood-brain barrier and therefore become potential drugs is still an open issue. One has to understand that we used a transition-state analogue against β-secretase and hence it was imperative, in our case, to target the inhibitor to endosomes where the enzyme was active. However, the reduction in dimensionality and subcellular targeting can now also be used as principles to design strategies to develop drugs against other membrane protein targets that are active at the plasma membrane and/or in intracellular compartments.

Acknowledgements The authors thank Hermann-Josef Kaiser, Robin Klemm and Unal Cokun for helpful discussions. L.R was supported by the Max-Planck Fellowship and the Alzheimer Forschungs Initiative e.V.

References

Adam G, Delbruck M (1968) Reduction of dimensionality in biological diffusion processes. San Francisco, W. H. Freeman and Company.

Andersen OM, Reiche J, Schmidt V, Gotthardt M, Spoelgen R, Behlke J, von Arnim CA, Breiderhoff T, Jansen P, Wu X, Bales KR, Cappai R, Masters CL, Gliemann J, Mufson EJ, Hyman BT, Paul SM, Nykjaer A, Willnow TE (2005) Neuronal sorting protein-related receptor sorLA/LR11 regulates processing of the amyloid precursor protein. Proc Natl Acad Sci USA 102: 13461–13466.

Annaert W, De Strooper, B (2002) A cell biological perspective on Alzheimer's disease. Annu Rev Cell Dev Biol 18: 25–51.

Cirrito JR, Kang JE, Lee J, Stewart FR, Verges DK, Silverio LM, Bu G, Mennerick S, Holtzman DM (2008) Endocytosis is required for synaptic activity-dependent release of amyloid-beta in vivo. Neuron 58: 42–51.

Cordy JM, Hussain I, Dingwall C, Hooper NM, Turner AJ (2003) Exclusively targeting beta-secretase to lipid rafts by GPI-anchor addition up-regulates beta-site processing of the amyloid precursor protein. Proc Natl Acad Sci USA 100: 11735–11740.

Daugherty BL, Green SA (2001) Endosomal sorting of amyloid precursor protein-P-selectin chimeras influences secretase processing. Traffic 2: 908–916.

de Gassart A, Geminard C, Hoekstra D,Vidal M (2004) Exosome secretion: the art of reutilizing nonrecycled proteins? Traffic 5: 896–903.

de Renzis S, Sonnichsen B, Zerial M (2002) Divalent Rab effectors regulate the sub-compartmental organization and sorting of early endosomes. Nature Cell Biol 4: 124–133.

De Strooper B (2003) Aph-1, Pen-2, and Nicastrin with Presenilin generate an active gamma-Secretase complex. Neuron 38: 9–12.

Edbauer D, Winkler E, Regula J., Pesold B, Steiner H, Haass C (2003) Reconstitution of gamma-secretase activity. Nature Cell Biol 5: 486–488.

Ehehalt R, Keller P, Haass C, Thiele C, Simons K (2003) Amyloidogenic processing of the Alzheimer beta-amyloid precursor protein depends on lipid rafts. J Cell Biol 160: 113–123.

Fevrier B, Vilette D, Archer F, Loew D, Faigle W, Vidal M, Laude H, Raposo G (2004) Cells release prions in association with exosomes. Proc Natl Acad Sci USA 101: 9683–9688.

Hayashi H, Kimura N, Yamaguchi H, Hasegawa K, Yokoseki T, Shibata M, Yamamoto N, Michikawa M, Yoshikawa Y, Terao K, et al. (2004) A seed for Alzheimer amyloid in the brain. J Neurosci 24: 4894–4902.

He X, Li F, Chang WP,Tang J (2005) GGA proteins mediate the recycling pathway of memapsin 2 (BACE). J Biol Chem 280: 11696–11703.

Hong L, Koelsch G, Lin X, Wu S, Terzyan S, Ghosh A K, Zhang XC, Tang J (2000) Structure of the protease domain of memapsin 2 (beta-secretase) complexed with inhibitor. Science 290: 150–153.

Hooper NM (2005) Roles of proteolysis and lipid rafts in the processing of the amyloid precursor protein and prion protein. Biochem Soc Trans 33: 335–338.

Kaether C, Schmitt S, Willem M, Haass C (2006) Amyloid precursor protein and notch intracellular domains are generated after transport of their precursors to the cell surface. Traffic 7: 408–415.

Kalvodova L, Kahya N, Schwille P, Ehehalt R, Verkade P, Drechsel D, Simons K (2005) Lipids as modulators of proteolytic activity of BACE: involvement of cholesterol, glycosphingolipids, and anionic phospholipids in vitro. J Biol Chem 280: 36815–36823.

Kinoshita A, Fukumoto H, Shah T, Whelan CM, Irizarry MC, Hyman BT (2003) Demonstration by FRET of BACE interaction with the amyloid precursor protein at the cell surface and in early endosomes. J Cell Sci 116: 3339–3346.

Knobloch M, Konietzko U, Krebs DC, Nitsch RM (2007) Intracellular Abeta and cognitive deficits precede beta-amyloid deposition in transgenic arcAbeta mice. Neurobiol Aging 28: 1297–1306.

Kojro E, Fahrenhol F (2005) The non-amyloidogenic pathway: structure and function of alpha-secretases. Subcell Biochem 38: 105–127.

Kojro E, Gimpl G, Lammich S, Marz W, Fahrenholz F (2001) Low cholesterol stimulates the nonamyloidogenic pathway by its effect on the alpha -secretase ADAM 10. Proc Natl Acad Sci USA 98: 5815–5820.

Lee SJ, Liyanage U, Bickel PE, Xia W, Lansbury PT Jr., Kosik KS (1998) A detergent-insoluble membrane compartment contains A beta in vivo. Nature Med 4: 730–734.

Pelchen-Matthews A, Raposo G, Marsh M (2004) Endosomes, exosomes and Trojan viruses. Trends Microbiol 12: 310–316.

Rajendran L, Simons K (2005) Lipid rafts and membrane dynamics. J Cell Sci 118: 1099–1102.

Rajendran L, Honsho M, Zahn TR, Keller P, Geiger KD, Verkade P, Simons K (2006) Alzheimer's disease beta-amyloid peptides are released in association with exosomes. Proc Natl Acad Sci USA 103: 11172–11177.

Rajendran L, Knobloch M, Geiger KD, Dienel S, Nitsch R, Simons K, Konietzko U (2007) Increased Abeta production leads to intracellular accumulation of Abeta in flotillin-1-positive endosomes. Neurodegener Dis 4: 164–170.

Rajendran L, Schneider A, Schlechtingen G, Weidlich S, Ries J, Braxmeier T, Schwille P, Schulz JB, Schroeder C, Simons M, Jennings G, Knölker HJ, Simons K (2008) Efficient inhibition of the Alzheimer's disease beta-secretase by membrane targeting. Science 320: 520–523.

Raposo G, Marks MS (2007) Melanosomes–dark organelles enlighten endosomal membrane transport. Nature Rev Mol Cell Biol 8: 786–797.

Raposo G, Nijman HW, Stoorvogel W, Liejendekker R, Harding CV, Melief CJ, Geuze HJ (1996) B lymphocytes secrete antigen-presenting vesicles. J Exp Med 183: 1161–1172.
Refolo LM, Sambamurti K, Efthimiopoulos S, Pappolla MA, Robakis NK (1995) Evidence that secretase cleavage of cell surface Alzheimer amyloid precursor occurs after normal endocytic internalization. J Neurosci Res 40: 694–706.
Resh MD (2004) Membrane targeting of lipid modified signal transduction proteins. Subcell Biochem 37: 217–232.
Rink J, Ghigo E, Kalaidzidis Y, Zerial M (2005). Rab conversion as a mechanism of progression from early to late endosomes. Cell 122: 735–749.
Rogaeva, E, Meng Y, Lee JH, Gu Y, Kawarai T, Zou F, Katayama T, Baldwin CT, Cheng R, Hasegawa H et al. (2007). The neuronal sortilin-related receptor SORL1 is genetically associated with Alzheimer disease. Nature Genet 39: 168–177.
Schneider A, Rajendran L, Honsho M, Gralle M, Donnert G, Wouters F, Hell S W, Simons M (2008) Flotillin-dependent clustering of the amyloid precursor protein regulates its endocytosis and amyloidogenic processing in neurons. J Neurosci 28: 2874–2882.
Selkoe DJ, Yamazaki T, Citron M, Podlisny MB, Koo EH, Teplow DB, Haass C (1996) The role of APP processing and trafficking pathways in the formation of amyloid beta-protein. Ann N Y Acad Sci 777: 57–64.
Simons M, Keller P, Dichgans J, Schulz JB (2001) Cholesterol and Alzheimer's disease: is there a link? Neurology 57: 1089–1093.
Small SA, Gandy S (2006) Sorting through the cell biology of Alzheimer's disease: intracellular pathways to pathogenesis. Neuron 52: 15–31.
Takahashi RH, Milner TA, Li F, Nam EE, Edgar MA, Yamaguchi H, Beal MF, Xu H, Greengard P, Gouras GK (2002) Intraneuronal Alzheimer abeta42 accumulates in multivesicular bodies and is associated with synaptic pathology. Am J Pathol 161: 1869–1879.
Tesco G, Koh YH, Kang E L, Cameron A N, Das S, Sena-Esteves M, Hiltunen M, Yang SH, Zhong Z, Shen Y, Simpkins JW, Tanzi RE (2007) Depletion of GGA3 stabilizes BACE and enhances beta-secretase activity. Neuron 54: 721–737.
Thery C, Boussac M, Veron P, Ricciardi-Castagnoli P, Raposo G, Garin J,Amigorena S (2001) Proteomic analysis of dendritic cell-derived exosomes: a secreted subcellular compartment distinct from apoptotic vesicles. J Immunol 166: 7309–7318.
Trajkovic K, Hsu C, Chiantia S, Rajendran L, Wenzel D, Wieland F, Schwille P, Brugger B, Simons M (2008) Ceramide triggers budding of exosome vesicles into multivesicular endosomes. Science 319: 1244–1247.
Tung JS, Davis DL, Anderson JP, Walker DE, Mamo S, Jewett N, Hom RK, Sinha S, Thorsett ED, John V (2002) Design of substrate-based inhibitors of human beta-secretase. J Med Chem 45: 259–262.
Vassar R (2002) Beta-secretase (BACE) as a drug target for Alzheimer's disease. Adv Drug Deliv Rev 54: 1589–1602.
Vetrivel KS, Cheng H, Lin W, Sakurai T, Li T, Nukina N, Wong PC, Xu H, Thinakaran G (2004) Association of gamma-secretase with lipid rafts in post-Golgi and endosome membranes. J Biol Chem 279: 44945–44954.
Yuyama K, Yamamoto N, Yanagisawa K (2008) Accelerated release of exosome-associated GM1 ganglioside (GM1) by endocytic pathway abnormality: another putative pathway for GM1-induced amyloid fibril formation. J Neurochem 105: 217–224.

Huntington's Disease: Function and Dysfunction of Huntingtin in Axonal Transport

Frédéric Saudou(✉) and Sandrine Humbert

Abstract Huntington's disease is a neurodegenerative disorder characterized by the dysfunction and death of striatal neurons in the brain. The mutation that causes Huntington's disease is an abnormal polyglutamine (polyQ) expansion in the huntingtin protein. One key, early pathogenic event in the disease is the alteration of axonal transport. This alteration is linked, at least in part, to a defect in huntingtin function in transport. Huntingtin is found on microtubules and is associated with proteins of the molecular motor machinery, including dynein and huntingtin-associated protein-1 (HAP1), which interacts with p150^{Glued} and kinesin-1. Wild-type huntingtin promotes the vesicular transport along microtubules. In contrast, the abnormal expansion in polyQ-huntingtin alters the HAP1-p150^{Glued} complex, leading to the molecular motors being depleted from the microtubules and to a decreased transport. Given the importance of transport in the nervous system, reduced transport leads to neuronal toxicity, and strategies to restore axonal transport could therefore be of therapeutic interest in Huntington's disease.

1 Introduction

Huntington's disease (HD) is a fatal neurodegenerative disorder that affects 1 in 10,000 to 25,000 individuals of European origin (Martin and Gusella 1986; Borrell-Pages et al. 2006a). HD is characterized by uncontrolled movements (chorea), personality changes, and dementia and causes the death of patients within 10 to 20 years of the appearance of the first clinical symptoms. Pathologically, it causes the specific dysfunction and death of neurons from the striatum. The defective IT15 gene in HD contains a trinucleotide CAG repeat expansion within its coding region that expresses a polyglutamine (polyQ) repeat in the N-terminus of

F. Saudou
Institut Curie – CNRS UMR 146, Bldg 110, Centre Universitaire, 91405 Orsay Cedex, France
E-mail: Frederic.Saudou@curie.fr

P. St. George-Hyslop et al. (eds.) *Intracellular Traffic and Neurodegenerative Disorders*, Research and Perspectives in Alzheimer's Disease,

the 350 kD protein huntingtin (HDCRG 1993; Snell et al. 1993). When the number of glutamine repeats exceeds 36, the gene encodes a version of huntingtin that leads to the disease. Although the mechanisms that cause disease are not fully understood, several studies have revealed a series of events that may ultimately lead to neuronal death in the brain.

Huntingtin is an indispensable protein that has anti-apoptotic properties. The protein is widely expressed in all tissues, with the highest levels being found in the testis and brain (DiFiglia et al. 1995; Gutekunst et al. 1995; Trottier et al. 1995). Studies in huntingtin knock-out mice have shown that huntingtin is required for normal embryonic development and neurogenesis: mice lacking huntingtin show extensive embryonic ectoderm cell death at E7.5 (Duyao et al. 1995; Nasir et al. 1995; Zeitlin et al. 1995; White et al. 1997). Huntingtin also plays an essential role postnatally, as the inactivation of the gene in the brain in adults leads to neurodegeneration (Dragatsis et al. 2000). Furthermore, the wild-type protein protects against polyQ-huntingtin-induced cell death in vivo and against neurodegeneration after ischemia (Rigamonti et al. 2000; Cattaneo et al. 2001; Ho et al. 2001; Leavitt et al. 2001; Zhang et al. 2003b). Huntingtin overexpression also increases the survival of serum-deprived or 3 nitropropionic acid (3-NP)-treated striatal cells (Rigamonti et al. 2001). Finally, the anti-apoptotic effect of huntingtin is supported by the observation that huntingtin downregulates activation of the procaspase 8 apoptotic pathway by sequestering HIP-1 (Hackam et al. 2000; Gervais et al. 2002). In contrast, when huntingtin contains an abnormal polyQ expansion, it becomes toxic. It induces the formation of neuritic and intranuclear inclusions, dysfunction of neurons and finally their death. The precise mechanisms underlying these phenomena are hardly understood, and how increased neuronal death in the brain relates to huntingtin function and dysfunction is still under debate.

We review here the newly discovered function of huntingtin in intracellular transport along microtubules. A better understanding of huntingtin biology has allowed the emergence of new concepts for the disease. First, neuronal dysfunction plays an important role in the appearance and progression of the clinical symptoms. HD should thus not simply be considered a disease of neuronal death. Second, it becomes clear that both the gain of a new toxic function of the mutant protein and the loss of the protective functions of wild-type huntingtin participate in the disease mechanisms that ultimately lead to the death of neurons in the brain.

2 Huntingtin and intracellular transport

Transport efficiency is of particular importance in neurons. To allow efficient communication between cell bodies and axon termini, molecular motor proteins continuously shuttle vesicles and organelles. The axonal transport process mostly involves microtubules and molecular motors that are considered to be unidirectional.

Dynein complexes are connected to retrograde transport, whereas kinesins are connected to anterograde transport.

Within cells, huntingtin is found in the cytoplasm, in neurites and, in particular, on microtubules (MTs; Gutekunst et al. 1995; Engelender et al. 1997; Gauthier et al. 2004). Huntingtin directly interacts with ß-tubulin (Hoffner et al. 2002). It associates with proteins of the molecular motor machinery, such as dynein and the huntingtin-associated protein-1 (HAP1). HAP1 itself associates with the p150^{Glued} subunit of dynactin and to the kinesin light chain 2, a subunit of kinesin-1 complex. In addition to the indirect association of huntingtin with dynactin through HAP1, huntingtin also interacts with the dynein intermediate chain of the dynein complex (Gutekunst et al. 1995; Engelender et al. 1997; Li et al. 1998; Gauthier et al. 2004; McGuire et al. 2006; Caviston et al., 2007).

Huntingtin is found colocalizing with vesicles including those containing brain-derived neurotrophic factor (BDNF; Gauthier et al. 2004). BDNF is particularly important in HD. Indeed, it is produced in the cortex and is transported to the striatum, the major site of degeneration in HD, where it supports neuronal differentiation and survival (Altar et al. 1997; Saudou et al. 1998; Zuccato et al. 2001; Baquet et al. 2004; Gauthier et al. 2004). BDNF inhibits polyQ-huntingtin-induced neuronal death and its level is abnormally low in HD patients. BDNF is synthesized from the large precursor protein pre-pro-BDNF, which is proteolytically processed and moves through the Golgi apparatus to the trans-Golgi network, where it is packaged into vesicles (Thomas and Davies 2005). BDNF-containing vesicles are then transported along MTs to the plasma membrane and subsequently released through the regulated secretory pathway. BDNF-containing vesicles are immunopositive for the classical markers of secretion, and their activity-dependent release requires an intact MT network, as it is blocked by nocodazole, a MT-depolymerizing agent (Gauthier et al. 2004).

Using BDNF as a marker of intracellular trafficking in cells, we showed that expression of huntingtin in neuroblastoma cell lines and in neurons enhances the velocity of BDNF-containing vesicles while reducing the percentage of time they spent pausing (Gauthier et al. 2004). In support of a positive role of huntingtin in stimulating axonal transport of vesicles that contain BDNF, downregulation of huntingtin by RNAi approaches leads to a decrease in the velocity of moving vesicles and an increase in the percentage of time spent pausing. These results are in agreement with *Drosophila* studies showing that a reduction in huntingtin protein level by RNAi approach results in axonal transport defects in larval nerves and neurodegeneration in adult eyes (Gunawardena et al. 2003). This huntingtin-dependent transport of BDNF vesicles along MTs is bidirectional, as huntingtin stimulates both anterograde and retrograde transport in axons (Gauthier et al. 2004; Dompierre et al. 2007).

The stimulatory effect of huntingtin on intracellular transport involves the direct interaction of huntingtin with dynein intermediate chain and with HAP1, which interacts with the p150^{Glued} subunit of dynactin and kinesin (Gauthier et al. 2004). Furthermore, short N-terminal fragments of huntingtin that do not contain the HAP1-interacting region are unable to stimulate intracellular transport. Also, BDNF transport is reduced after downregulation of HAP1 protein and, under these conditions, huntingtin is unable to enhance BDNF trafficking.

Although these studies revealed a role for huntingtin in axonal transport, more work needs to be done to establish the extent to which axonal transport depends on huntingtin. Huntingtin stimulates the dynamic of BDNF-containing vesicles; however, whether other types of vesicles are regulated by huntingtin remains to be established. The observations that transport of amyloid precursor protein vesicles depends on HAP1 and that downregulation of huntingtin alters the general axonal transport in *Drosophila* strongly suggest that huntingtin regulates the transport of other small vesicles (Gunawardena et al. 2003; McGuire et al. 2006). Milton, a *Drosophila* ortholog of HAP1, participates in the axonal transport of mitochondria, thus raising the possibility that huntingtin and HAP1 could also regulate the transport of mitochondria (Glater et al. 2006). However, the velocity of mitochondria is not regulated by huntingtin and, whereas HAP1 overexpression leads to the redistribution of BDNF vesicles in cells, it has no effect on mitochondria. Conversely, Milton, though known to redistribute mitochondria in cells (Stowers et al. 2002), has no effect on BDNF vesicles (Gauthier et al. 2004 and data not shown). Therefore, HAP1 and Milton show specificity in the type of cargoes they are transporting.

3 Transport Deficit in HD

In disease, the presence of an abnormal polyQ expansion in huntingtin leads to a loss of the stimulatory function of huntingtin in transport (Gauthier et al. 2004). Indeed, while expressing wild-type huntingtin increases the mean velocity of BDNF vesicles, polyQ-huntingtin has no effect, and cells homozygous for mutant huntingtin show a reduced BDNF transport. The physiological consequence of an altered transport is a reduced BDNF support and a higher susceptibility of striatal neurons to death. Furthermore, BDNF levels are reduced in the striatum of HD patients.

What are the underlying molecular mechanisms? When huntingtin contains the pathological polyQ expansion, it interacts more strongly with HAP1 and $p150^{Glued}$ (Li et al. 1995; Li et al. 1998; Gauthier et al. 2004), which directly modifies the huntingtin/HAP1/$p150^{Glued}$ complex, as revealed by sucrose gradient fractionation and immunoprecipitation experiments. As a result, the molecular motors detach from the MTs and the processivity of vesicles along the MTs is reduced. As discussed earlier, a reduction in huntingtin levels or in the expression of mutant huntingtin reduces transport (Gunawardena et al. 2003; Gauthier et al. 2004). Therefore, in early stages of HD, the disruption of huntingtin (soluble form)/HAP1 interaction causes huntingtin to no longer play a role in transport.

In later stages of the disease, in addition to nuclear aggregation, N-terminal huntingtin fragments form aggregates that accumulate in axonal processes and terminals (Li et al. 2000). N-terminal huntingtin polypeptide fragments containing the polyQ expansion cause axonal transport defects in cellular and *Drosophila* models of HD (Li et al. 2000; Gunawardena et al. 2003; Szebenyi et al. 2003; Lee et al. 2004; Trushina et al. 2004; Orr et al. 2008). These defects subsequently participate in neuronal death. Aggregation is involved in alterating axonal transport, with aggregated polyQ-proteins accumulating in axons and titrating motor proteins, particularly p150^{Glued} and kinesin heavy chain (KHC), from other cargoes and pathways. These aggregates also physically block the circulating vesicles or organelles such as mitochondria (Orr et al. 2008).

4 Rescuing the Deficient BDNF Dynamics as a Therapeutic Approach

In the case of HD, as huntingtin directly controls transport of the pro-survival factor BDNF, enhancing transport or more generally rescuing the defective BDNF dynamics might be a promising therapeutic approach. In this regard, we identified two pathways of interest.

Cystamine, a compound described as a transglutaminase (TGase) inhibitor, is one of the few candidate drugs being considered for the treatment of HD, as it is neuroprotective in several HD mice models (Dedeoglu et al. 2002; Karpuj et al. 2002; Mastroberardino et al. 2002; Wang et al. 2005). TGase is a calcium-dependent enzyme that catalyzes the formation of ε (α-glutamyl)lysine isopeptide bonds between a polypeptide-bound glutamine and a lysine of the protein substrate (Melino and Piacentini 1998; Lesort et al. 2000). Given their enzymatic properties, TGases might promote aggregate formation in HD. However, cystamine treatment of HD mice does not necessarily result in fewer neuronal intranuclear inclusions (NIIs) (Karpuj et al. 2002), and an increase in NIIs is observed in HD mice that are deficient for one of the TGase isoenzymes, tissue transglutaminase 2 (TGase 2; Mastroberardino et al. 2002; Bailey and Johnson 2005). Thus the mechanims by which cystamine is neuroprotective in HD are unclear. We demonstrated that part of the neuroprotective effect of cystamine is due to its promotion of secretion of BDNF (Borrell-Pages et al. 2006b). Cystamine has two quite distinct actions to induce BDNF secretion. First, it increases the steady-state levels of the heat shock protein, HSJ1b mRNA, which stimulates the secretory pathway through its action on clathrin-coated vesicle formation, and second, it inhibits transglutaminase, which has a negative effect on BDNF sorting. Interestingly, we also showed that cysteamine, the FDA-approved reduced form of cystamine, is neuroprotective in HD mice by enhancing BDNF levels in brain.

Among other molecules of therapeutic interest are HDAC inhibitors such as suberoylanilide hydroxamic acid (SAHA) and trichostatin A (TSA), which have shown neuroprotective effects by inhibiting the HDAC1 enzyme (Butler and Bates 2006). These drugs are not specific for a given HDAC but also act on other HDACs, such as HDAC6 (Haggarty et al. 2003). Unlike other histone deacetylases, HDAC6 is a cytoplasmic enzyme that interacts with and deacetylates MTs in vitro and in vivo (Hubbert et al. 2002; Matsuyama et al. 2002; Zhang et al. 2003a). We demonstrated that HDAC inhibitors that selectively enhance tubulin but not histone acetylation lead to the stimulation of MT-dependent transport of BDNF and prevent the alteration observed in HD mutant cells (Dompierre et al. 2007). This effect is specific to HDAC6 inhibition and to the acetylation of α-tubulin at lysine 40. Using in vitro experiments, we showed that purified cytoplasmic dynein and recombinant kinesin-1 bound more effectively to acetylated MTs. Enhancing MT acetylation led to the recruitment of molecular motors kinesin-1 and cytoplasmic dynein to MTs, thereby stimulating anterograde and retrograde transport. As a consequence, this increased transport enhanced the anterograde flux of vesicles and the subsequent release of BDNF in normal and pathological conditions.

Therefore, by stimulating BDNF secretion from the Golgi to the cytoplasm or by directly targeting the microtubules, BDNF dynamics are stimulated and the deficit observed in HD is rescued. As stated above, BDNF is depleted in HD human brains (Gauthier et al. 2004). In mouse models of HD, BDNF brain and blood levels are low and can be increased by injection of cysteamine (Borrell-Pages et al. 2006b). Similarly, in primate HD models, the low levels of BDNF in blood can be increased by cysteamine treatment, suggesting that blood BDNF could be used to follow disease progression and validate the neuroprotective effects of drugs aiming to restore the defective intracellular vesicular dynamics in HD.

5 Conclusion

Since the discovery of the abnormal polyglutamine expansion in huntingtin as the dominant mutation responsible for HD, most studies in the field have focused on understanding the gain of the toxic function elicited by this mutation. The role of huntingtin as a stimulator of BDNF intracellular transport is in agreement with findings that huntingtin possesses anti-apoptotic properties, which are also originally linked to an increased BDNF transcriptional activity (Zuccato et al., 2001). These anti-apoptotic properties are lost in disease and also contribute to pathogenesis, further supporting the importance of studying normal huntingtin function. Furthermore, understanding normal huntingtin function not only leads to the discovery of new pathways of pathological importance but also outlines new therapeutic targets. In the case of axonal transport, compounds such as cystamine/cysteamine and HDACs inhibitors that enhance intracelular dynamics are of therapeutic interest. This approach is of utmost importance, as no treatment currently exists for this devastating disorder.

References

Altar CA, Cai N, Bliven T, Juhasz M, Conner JM, Acheson AL, Lindsay RM, Wiegand SJ (1997) Anterograde transport of brain-derived neurotrophic factor and its role in the brain. Nature 389:856–860.

Bailey CD, Johnson GV (2005) Tissue transglutaminase contributes to disease progression in the R6/2 Huntington's disease mouse model via aggregate-independent mechanisms. J Neurochem 92:83–92.

Baquet ZC, Gorski JA, Jones KR (2004) Early striatal dendrite deficits followed by neuron loss with advanced age in the absence of anterograde cortical brain-derived neurotrophic factor. J Neurosci 24:4250–4258.

Borrell-Pages M, Zala D, Humbert S, Saudou F (2006a) Huntington's disease: from huntingtin function and dysfunction to therapeutic strategies. Cell Mol Life Sci 63:2642–2660.

Borrell-Pages M, Canals JM, Cordelieres FP, Parker JA, Pineda JR, Grange G, Bryson EA, Guillermier M, Hirsch E, Hantraye P, Cheetham ME, Neri C, Alberch J, Brouillet E, Saudou F, Humbert S (2006b) Cystamine and cysteamine increase brain levels of BDNF in Huntington disease via HSJ1b and transglutaminase. J Clin Invest 116:1410–1424.

Butler R, Bates GP (2006) Histone deacetylase inhibitors as therapeutics for polyglutamine disorders. Nature Rev Neurosci 7:784–796.

Cattaneo E, Rigamonti D, Goffredo D, Zuccato C, Squitieri F, Sipione S (2001) Loss of normal huntingtin function: new developments in Huntington's disease research. Trends Neurosci 24:182–188.

Caviston JP, Ross JL, Antony SM, Tokito M, Holzbaur EL (2007) Huntingtin facilitates dynein/dynactin-mediated vesicle transport. Proc Natl Acad Sci USA 104:10045–10050.

Dedeoglu A, Kubilus JK, Jeitner TM, Matson SA, Bogdanov M, Kowall NW, Matson WR, Cooper AJ, Ratan RR, Beal MF, Hersch SM, Ferrante RJ (2002) Therapeutic effects of cystamine in a murine model of Huntington's disease. J Neurosci 22:8942–8950.

DiFiglia M, Sapp E, Chase K, Schwarz C, Meloni A, Young C, Martin E, Vonsattel JP, Carraway R, Reeves SA, Boyce FM, Aronin N (1995) Huntingtin is a cytoplasmic protein associated with vesicles in human and rat brain neurons. Neuron 14:1075–1081.

Dompierre JP, Godin JD, Charrin BC, Cordelieres FP, King SJ, Humbert S, Saudou F (2007) Histone deacetylase 6 inhibition compensates for the transport deficit in Huntington's disease by increasing tubulin acetylation. J Neurosci 27:3571–3583.

Dragatsis I, Levine MS, Zeitlin S (2000) Inactivation of hdh in the brain and testis results in progressive neurodegeneration and sterility in mice. Nature Genet 26:300–306.

Duyao MP, Auerbach AB, Ryan A, Persichetti F, Barnes GT, McNeil SM, Ge P, Vonsattel JP, Gusella JF, Joyner AL, MacDonald ME (1995) Inactivation of the mouse Huntington's disease gene homolog Hdh. Science 269:407–410.

Engelender S, Sharp AH, Colomer V, Tokito MK, Lanahan A, Worley P, Holzbaur EL, Ross CA (1997) Huntingtin-associated protein 1 (HAP1) interacts with the p150Glued subunit of dynactin. Human Mol Genet 6:2205–2212.

Gauthier LR, Charrin BC, Borrell-Pages M, Dompierre JP, Rangone H, Cordelieres FP, De Mey J, MacDonald ME, Lessmann V, Humbert S, Saudou F (2004) Huntingtin controls neurotrophic support and survival of neurons by enhancing BDNF vesicular transport along microtubules. Cell 118:127–138.

Gervais FG, Singaraja R, Xanthoudakis S, Gutekunst CA, Leavitt BR, Metzler M, Hackam AS, Tam J, Vaillancourt JP, Houtzager V, Rasper DM, Roy S, Hayden MR, Nicholson DW (2002) Recruitment and activation of caspase-8 by the Huntingtin-interacting protein Hip-1 and a novel partner Hippi. Nature Cell Biol 4:95–105.

Glater EE, Megeath LJ, Stowers RS, Schwarz TL (2006) Axonal transport of mitochondria requires milton to recruit kinesin heavy chain and is light chain independent. J Cell Biol 173:545–557.

Gunawardena S, Her LS, Brusch RG, Laymon RA, Niesman IR, Gordesky-Gold B, Sintasath L, Bonini NM, Goldstein LS (2003) Disruption of axonal transport by loss of huntingtin or expression of pathogenic polyQ proteins in Drosophila. Neuron 40:25–40.

Gutekunst CA, Levey AI, Heilman CJ, Whaley WL, Yi H, Nash NR, Rees HD, Madden JJ, Hersch SM (1995) Identification and localization of huntingtin in brain and human lymphoblastoid cell lines with anti-fusion protein antibodies. Proc Natl Acad Sci USA 92:8710–8714.

Hackam AS, Yassa AS, Singaraja R, Metzler M, Gutekunst CA, Gan L, Warby S, Wellington CL, Vaillancourt J, Chen N, Gervais FG, Raymond L, Nicholson DW, Hayden MR (2000) Huntingtin interacting protein 1 (HIP-1) induces apoptosis via a novel caspase-dependent death effector domain. J Biol Chem 275:41299–41308.

Haggarty SJ, Koeller KM, Wong JC, Grozinger CM, Schreiber SL (2003) Domain-selective small-molecule inhibitor of histone deacetylase 6 (HDAC6)-mediated tubulin deacetylation. Proc Natl Acad Sci USA 100:4389–4394.

HDCRG (1993) A novel gene containing a trinucleotide repeat that is expanded and unstable on Huntington's disease chromosomes. Cell 72:971–983.

Ho LW, Brown R, Maxwell M, Wyttenbach A, Rubinsztein DC (2001) Wild type Huntingtin reduces the cellular toxicity of mutant Huntingtin in mammalian cell models of Huntington's disease. J Med Genet 38:450–452.

Hoffner G, Kahlem P, Djian P (2002) Perinuclear localization of huntingtin as a consequence of its binding to microtubules through an interaction with beta-tubulin: relevance to Huntington's disease. J Cell Sci 115:941–948.

Hubbert C, Guardiola A, Shao R, Kawaguchi Y, Ito A, Nixon A, Yoshida M, Wang XF, Yao TP (2002) HDAC6 is a microtubule-associated deacetylase. Nature 417:455–458.

Karpuj MV, Becher MW, Springer JE, Chabas D, Youssef S, Pedotti R, Mitchell D, Steinman L (2002) Prolonged survival and decreased abnormal movements in transgenic model of Huntington disease, with administration of the transglutaminase inhibitor cystamine. Nature Med 8:143–149.

Leavitt BR, Guttman JA, Hodgson JG, Kimel GH, Singaraja R, Vogl AW, Hayden MR (2001) Wild-type huntingtin reduces the cellular toxicity of mutant huntingtin in vivo. Am J Human Genet 68:313–324.

Lee WC, Yoshihara M, Littleton JT (2004) Cytoplasmic aggregates trap polyglutamine-containing proteins and block axonal transport in a Drosophila model of Huntington's disease. Proc Natl Acad Sci USA 101:3224–3229.

Lesort M, Tucholski J, Miller ML, Johnson GV (2000) Tissue transglutaminase: a possible role in neurodegenerative diseases. Prog Neurobiol 61:439–463.

Li H, Li SH, Johnston H, Shelbourne PF, Li XJ (2000) Amino-terminal fragments of mutant huntingtin show selective accumulation in striatal neurons and synaptic toxicity. Nature Genet 25:385–389.

Li SH, Gutekunst CA, Hersch SM, Li XJ (1998) Interaction of huntingtin-associated protein with dynactin P150Glued. J Neurosci 18:1261–1269.

Li XJ, Li SH, Sharp AH, Nucifora FC, Jr., Schilling G, Lanahan A, Worley P, Snyder SH, Ross CA (1995) A huntingtin-associated protein enriched in brain with implications for pathology. Nature 378:398–402.

Martin JB, Gusella JF (1986) Huntington's disease. Pathogenesis and management. New Engl J Med 315:1267–1276.

Mastroberardino PG, Iannicola C, Nardacci R, Bernassola F, De Laurenzi V, Melino G, Moreno S, Pavone F, Oliverio S, Fesus L, Piacentini M (2002) 'Tissue' transglutaminase ablation reduces neuronal death and prolongs survival in a mouse model of Huntington's disease. Cell Death Differ 9:873–880.

Matsuyama A, Shimazu T, Sumida Y, Saito A, Yoshimatsu Y, Seigneurin-Berny D, Osada H, Komatsu Y, Nishino N, Khochbin S, Horinouchi S, Yoshida M (2002) In vivo destabilization of dynamic microtubules by HDAC6-mediated deacetylation. Embo J 21:6820–6831.

McGuire JR, Rong J, Li SH, Li XJ (2006) Interaction of Huntingtin-associated protein-1 with kinesin light chain: implications in intracellular trafficking in neurons. J Biol Chem 281:3552–3559.

Melino G, Piacentini M (1998) 'Tissue' transglutaminase in cell death: a downstream or a multifunctional upstream effector? FEBS Lett 430:59–63.

Nasir J, Floresco SB, O'Kusky JR, Diewert VM, Richman JM, Zeisler J, Borowski A, Marth JD, Phillips AG, Hayden MR (1995) Targeted disruption of the Huntington's disease gene results in embryonic lethality and behavioral and morphological changes in heterozygotes. Cell 81:811–823.

Orr AL, Li S, Wang CE, Li H, Wang J, Rong J, Xu X, Mastroberardino PG, Greenamyre JT, Li XJ (2008) N-terminal mutant huntingtin associates with mitochondria and impairs mitochondrial trafficking. J Neurosci 28:2783–2792.

Rigamonti D, Bauer JH, De-Fraja C, Conti L, Sipione S, Sciorati C, Clementi E, Hackam A, Hayden MR, Li Y, Cooper JK, Ross CA, Govoni S, Vincenz C, Cattaneo E (2000) Wild-type huntingtin protects from apoptosis upstream of caspase-3. J Neurosci 20:3705–3713.

Rigamonti D, Sipione S, Goffredo D, Zuccato C, Fossale E, Cattaneo E (2001) Huntingtin's neuroprotective activity occurs via inhibition of procaspase-9 processing. J Biol Chem 276:14545–14548.

Saudou F, Finkbeiner S, Devys D, Greenberg ME (1998) Huntingtin acts in the nucleus to induce apoptosis but death does not correlate with the formation of intranuclear inclusions. Cell 95: 55–66.

Snell RG, MacMillan JC, Cheadle JP, Fenton I, Lazarou LP, Davies P, MacDonald ME, Gusella JF, Harper PS, Shaw DJ (1993) Relationship between trinucleotide repeat expansion and phenotypic variation in Huntington's disease. Nature Genet 4:393–397.

Stowers RS, Megeath LJ, Gorska-Andrzejak J, Meinertzhagen IA, Schwarz TL (2002) Axonal transport of mitochondria to synapses depends on milton, a novel Drosophila protein. Neuron 36:1063–1077.

Szebenyi G, Morfini GA, Babcock A, Gould M, Selkoe K, Stenoien DL, Young M, Faber PW, MacDonald ME, McPhaul MJ, Brady ST (2003) Neuropathogenic forms of huntingtin and androgen receptor inhibit fast axonal transport. Neuron 40:41–52.

Thomas K, Davies A (2005) Neurotrophins: a ticket to ride for BDNF. Curr Biol 15:R262–264.

Trottier Y, Devys D, Imbert G, Saudou F, An I, Lutz Y, Weber C, Agid Y, Hirsch EC, Mandel JL (1995) Cellular localization of the Huntington's disease protein and discrimination of the normal and mutated form. Nature Genet 10:104–110.

Trushina E, Dyer RB, Badger JD, 2nd, Ure D, Eide L, Tran DD, Vrieze BT, Legendre-Guillemin V, McPherson PS, Mandavilli BS, Van Houten B, Zeitlin S, McNiven M, Aebersold R, Hayden M, Parisi JE, Seeberg E, Dragatsis I, Doyle K, Bender A, Chacko C, McMurray CT (2004) Mutant huntingtin impairs axonal trafficking in mammalian neurons in vivo and in vitro. Mol Cell Biol 24:8195–8209.

Wang X, Sarkar A, Cicchetti F, Yu M, Zhu A, Jokivarsi K, Saint-Pierre M, Brownell AL (2005) Cerebral PET imaging and histological evidence of transglutaminase inhibitor cystamine induced neuroprotection in transgenic R6/2 mouse model of Huntington's disease. J Neurol Sci 231:57–66.

White JK, Auerbach W, Duyao MP, Vonsattel JP, Gusella JF, Joyner AL, MacDonald ME (1997) Huntingtin is required for neurogenesis and is not impaired by the Huntington's disease CAG expansion. Nature Genet 17:404–410.

Zeitlin S, Liu JP, Chapman DL, Papaioannou VE, Efstratiadis A (1995) Increased apoptosis and early embryonic lethality in mice nullizygous for the Huntington's disease gene homologue. Nature Genet 11:155–163.

Zhang Y, Li N, Caron C, Matthias G, Hess D, Khochbin S, Matthias P (2003a) HDAC-6 interacts with and deacetylates tubulin and microtubules in vivo. Embo J 22:1168–1179.

Zhang Y, Li M, Drozda M, Chen M, Ren S, Mejia Sanchez RO, Leavitt BR, Cattaneo E, Ferrante RJ, Hayden MR, Friedlander RM (2003b) Depletion of wild-type huntingtin in mouse models of neurologic diseases. J Neurochem 87:101–106.

Zuccato C, Ciammola A, Rigamonti D, Leavitt BR, Goffredo D, Conti L, MacDonald ME, Friedlander RM, Silani V, Hayden MR, Timmusk T, Sipione S, Cattaneo E (2001) Loss of huntingtin-mediated BDNF gene transcription in Huntington's disease. Science 293:493–498.

The Role of Retromer in Neurodegenerative Disease

Claire F. Skinner and Matthew N.J. Seaman(✉)

Abstract Bi-directional membrane traffic between the Golgi and endosomes plays a vital role in the biogenesis of lysosomes and the localisation of many membrane proteins with diverse physiological functions. The receptors that mediate sorting of lysosomal hydrolases at the Golgi traffic rapidly between the Golgi and endosomes to deliver newly synthesised hydrolases to a pre-lysosomal endosome before returning to the Golgi to repeat the process. The mislocalisation of endosomal and/or lysosomal proteins due to aberrant protein sorting can give rise to a range of pathologies, and there are emerging strands of evidence that defects in the endosome-to-Golgi retrieval pathway contribute significantly to neurodegenerative diseases such as Alzheimer's disease. The retromer complex that is conserved from yeast to humans plays a major role in endosomal protein sorting and is required for endosome-to-Golgi retrieval. In this review we will discuss the identification, assembly, membrane association and function of the retromer complex and will describe recent evidence linking retromer function with neurodegenerative disease.

1 Introduction

Biosynthetic transport of soluble hydrolases to the lysosome/vacuole is a receptor-mediated process conserved from simple eukaryotes like yeast to higher eukaryotes such as mammals. In the yeast *Saccharomyces cerevisiae*, Vps10p, a type I transmembrane protein binds hydrolases such as carboxypeptidase Y (CPY) in the late-Golgi and is sorted into vesicles for delivery to the prevacuolar compartment (PVC) by the Golgi-associated, γ-ear containing, ARF binding (GGA) proteins (Marcusson et al. 1994; Cereghino et al. 1995; Cooper and Stevens 1996; Costaguta

M.N.J. Seaman
Cambridge Institute for Medical Research/Dept. Clinical Biochemistry, University of Cambridge, Addenbrookes Hospital, Hills Road, Cambridge, CB2 0XY, UK, E-mail: mnjs100@cam.ac.uk

P. St. George-Hyslop et al. (eds.) *Intracellular Traffic and Neurodegenerative Disorders*, Research and Perspectives in Alzheimer's Disease,

et al. 2001). After delivery to the PVC, Vps10p and ligand dissociate, leaving the receptor free to be recycled back to the late-Golgi. This process is mirrored in mammalian cells with the exception that the receptor that sorts lysosomal hydrolases is the mannose-6-phosphate receptor (MPR). There are two distinct MPRs, the 46 kDa cation-dependent-MPR (CD-MPR), and the ~300 kDa cation-independent-MPR (CI-MPR), which share some homology in their respective lumenal domains (Kornfeld 1992). Both MPRs have acidic di-leucine motifs in their cytoplasmic tails that are recognised by the mammalian GGA proteins to mediate sorting of the MPRs into trans-Golgi-network (TGN)-derived, clathrin-coated vesicles for delivery to an endosomal compartment (Dell'Angelica et al. 2000; Shiba et al. 2002; Misra et al. 2002). As in yeast, the two MPRs traffic in a cyclical manner between the TGN and endosomes, thus maintaining the forward transport of newly synthesised hydrolases to the lysosome.

In mammals, lysosomes play a vital role in the degradation of endocytosed macromolecules (e.g., low-density lipoprotein – LDL), downregulation of activated tyrosine kinase receptors, antigen presentation, phagocytosis and autophagy, and there are many examples of genetic disease caused by mutations to the lysosomal hydrolases that perform degradative functions in lysosomes. These diseases are usually grouped together under the umbrella term "lysosomal storage disorders" (LSDs), but most share a common pathology of progressive neuronal degeneration. For example, Tay-Sachs disease results from loss of β-hexosaminidase function, which leads to an accumulation of GM2 gangliosides in the nervous system (Ni et al. 2006). Other lysosomal storage disorders result from mutations to membrane proteins; for example, Niemann-Pick type C disease, which results from mutations to the NPC1 gene, causes an accumulation of cholesterol in lysosomes (Sturley et al. 2004). Whilst it has been well established that sorting and delivery of lysosomal hydrolases have important roles to play in neurodegenerative disease such as LSDs, there has been relatively little attention directed towards the endosome-to-Golgi retrieval pathway and its importance in lysosome biogenesis and neurodegenerative disease. However, there is now compelling evidence that the function of the retromer complex in endosome-to-Golgi retrieval plays a significant role in sorting proteins involved in neurodegenerative disease.

2 Identification of the Retromer Complex

Genetic screens in yeast have proven invaluable in identifying the key participants in many membrane trafficking pathways, and transport between the Golgi and the vacuole (which is equivalent to the mammalian lysosome) is no exception. The *VPS10* gene is one of more than 60 vacuole protein-sorting genes discovered through the analysis of mutants that are defective in trafficking to the vacuole (Bryant and Stevens 1998). Detailed examination of the phenotype of the *vps10* mutants revealed that, whilst Vps10p is essential for transport of CPY to the vacuole, the receptor is apparently not necessary for transport of another soluble hydrolase, namely proteinase A (PrA; Westphal et al. 1996). This observation led to the search for other *vps* mutants that displayed a similar cargo-selective defect in vacuolar protein sorting, with the rationale being that mutants with a similar phenotype to *vps10* mutants would be likely to encode genes required for the trafficking/function of Vps10p.

Three *vps* genes were identified - *VPS29, VPS30* and *VPS35* - all of which had phenotypes similar to *vps10* mutants (Seaman et al. 1997). All three encoded soluble hydrophilic proteins that were able to associate peripherally with membranes. Deletion of any of the three genes resulted in Vps10p becoming mislocalised to the vacuolar membrane. Strikingly, it was observed that Vps35p would "follow" Vps10p to the vacuolar membrane in a *vps29Δ* mutant. These data hinted that Vps35p might be able to physically associate with Vps10p, presumably via the large cytoplasmic tail of Vps10p. This prediction was subsequently proved to be true through the analysis of mutant alleles of *vps35* and the use of biochemical experiments that demonstrated that Vps35p could interact directly with cargo proteins such as Vps10p (Nothwehr et al. 1999, 2000).

Epistasis experiments designed to determine the site of function of Vps35p suggested that Vps35p operated at the endosomal membrane. Loss of *VPS35* would, therefore, cause a defect in the retrieval of Vps10p from the endosome, resulting in mislocalisation to the vacuolar membrane. Once delivered to the vacuole, Vps10p had no means of escape and, consequently, there would be insufficient Vps10p in the late-Golgi to carry out the job of sorting newly synthesised CPY (Seaman et al. 1997). This hypothesis was supported by data from an experiment in a strain carrying a temperature-sensitive allele of *vps35p*. CPY sorting at the permissive temperature was normal, but upon shifting to the non-permissive temperature, there was a steady increase in the missorting and secretion of CPY over time. This phenotype coincided with the appearance of Vps10p in the vacuolar membrane fraction.

A direct interaction between Vps35p and Vps29p was demonstrated by co-immunoprecipitation following stabilisation by crosslinking. Three additional bands were observed in these crosslinking experiments, and these were subsequently identified as Vps5p, Vps17p and Vps26p (Seaman et al. 1998). Vps5p was localised to the endosome by both immunofluorescence and immuno-gold electron microscopy (EM) and was shown to be important for the proper localisation of Vps10p (Horazdovsky et al. 1997; Nothwehr and Hindes 1997). The complex of five Vps proteins - Vps5p, Vps17p, Vps26p, Vps29p and Vps35p - was subsequently named retromer to underline its role in retrieval from the endosome (Seaman et al. 1998). Interestingly, Vps30p was found to not be part of retromer. The role that Vps30p plays in endosome-to-Golgi retrieval is discussed in detail later.

The components of retromer are conserved through evolution (Renfrew-Haft et al. 2000). The homologues of yeast retromer were initially identified independently of each other. The mammalian VPS35 gene was identified as a **m**aternally **e**nriched **m**essenger RNA and was designated Mem3 (Hwang et al. 1996). The homologue of Vps26p was first cloned through a transgenic mouse screen for genes essential for embryogenesis and was named Hβ58 (Lee et al. 1992). A homologue of Vps29p was identified by searches of the EST database (Seaman et al. 1997), whereas the mammalian homologue of Vps5p was found to interact with the cytoplasmic tail of the epidermal growth factor receptor (EGFR) in a yeast two-hybrid screen and named sorting nexin-1 (SNX1; Kurten et al. 1996). The one exception to the conservation of yeast retromer is Vps17p, which does not have an obvious homologue in mammalian cells. SNX2 has been shown to associate with SNX1 and has, therefore, been proposed to function as the Vps17p ortholog (Renfrew-Haft et al. 2000).

3 Assembly of the Retromer Complex

Initial studies of retromer in yeast were directed towards understanding how the complex assembles and assigning functions to the individual components. By both biochemical and phenotypic criteria, retromer could be essentially dissected into two subcomplexes: the Vps35p/29p/26p complex and the Vps5p/17p complex (Seaman et al. 1998). Vps5p was identified as a binding partner of Vps17p independently of the discovery of retromer (Horazdovsky et al. 1997). The phenotypes of *vps5* and *vps17* mutants are virtually indistinguishable from each other, indicating that the two proteins are likely to interact (Kohrer and Emr 1993; Horazdovsky et al. 1997; Nothwehr and Hindes 1997). In fact, Vps5p/17p form a very stable dimer. Loss of Vps5p results in Vps17p becoming unstable and being degraded with a half-life of approximately 45 mins. Absence of Vps17p does not cause Vps5p to be degraded but results in Vps5p being unable to interact with the other components of retromer (Seaman and Williams 2002).

Retromer can be dissociated into its two subcomplexes by treatment with high salt buffers. The two subcomplexes can be separated from each other by gel filtration chromatography. Removal of the salt by dialysis allows the complex to reassemble (Seaman et al. 1998). Gel filtration and native immunoprecipitation studies indicated that Vps29p played a vital role within the complex, being necessary for the interaction between Vps35p and Vps5p/17p. Absence of the Vps29 protein did not, however, prevent the interaction between Vps35p and Vps26p. Loss of Vps26p did not have the same effect with the remaining four members of retromer being able to associate with each other, albeit rather weakly. Deletion of both *VPS29* and *VPS26* results in Vps35p becoming unstable, revealing that Vps35p requires interaction with either Vps29p or Vps26p for its correct folding (Seaman et al. 1998; Reddy and Seaman 2001).

Even though Vps35p/29p/26p are conserved in evolution, the primary structure of the Vps35, Vps29 and Vps26 proteins did not offer any clues regarding the function of any of these proteins, as there were no obvious domains of predicted secondary structure. In contrast, both Vps5p and Vps17p have regions predicted to form coiled-coils in their carboxyl-terminal halves. Vps5p and Vps17p are members of the sorting nexin (SNX) family of proteins. Defining features of SNX proteins are the phox homology (PX) domains. These are conserved lipid-binding domains that have been shown to be important for binding to phosphotidyl inositol 3-phosphate

(Ptd Ins-3P; Yu and Lemmon 2001). Vps5p and Vps17p both have PX domains and, therefore, could potentially bind Ptd Ins-3P. Therefore, by exploiting the different domain boundaries of Vps5p and Vps17p, the interactions of these proteins were studied in detail through the generation of various truncation constructs of both Vps5p and Vps17p. It was shown that Vps5p and Vps17p interact with each other through their respective carboxyl-terminal regions that have predicted coiled-coil domains. The amino-terminal domain of Vps5p is both necessary and sufficient to interact with the Vps35p/29p/26p subcomplex, but this interaction is facilitated in vivo through the interaction of Vps17p with Vps5p (Seaman and Williams 2002).

In addition to being able to interact with Vps17p and also the Vps35p/29p/26p complex, Vps5p displayed self-assembly activity in vitro. Recombinant Vps5p expressed in bacteria can assemble into large (>1.5 MDa), homogeneous, oligomeric particles that are 15–20 nm in diameter (Seaman et al. 1998). This self-assembly activity has also been observed for SNX1, the mammalian Vps5p homologue (Kurten et al. 2001). The physiological importance of this self-assembly activity is discussed in more detail later.

The data gathered in yeast have been complemented by studies on assembly of mammalian retromer using the yeast two-hybrid system. By this approach, the binding sites for hVPS26 and hVPS29 on hVPS35 have been mapped to the amino- and carboxyl-termini, respectively (Renfrew-Haft et al. 2000). Mutations to a conserved motif in Vps35p blocked the interaction with Vps26p and caused a dominant negative phenotype, whereas analogous mutations in the mammalian VPS35 resulted in VPS35 being unable to bind VPS26 and be recruited to the endosomal membrane (Gokool et al. 2007a; Zhao et al. 2007; Restrepo et al. 2007). The amino-terminus of hVPS35 also demonstrated an affinity for SNX1, the homologue of Vps5p (Renfrew-Haft et al. 2000). The assembly of retromer is shown schematically in Fig. 1. Further understanding of the assembly of retromer and the roles of the individual components has been accelerated by the determination of the crystal structure of retromer.

4 Insights from Structural Studies

The crystal structures of human and mouse VPS29 have been solved and, due to the high degree of identity between the two proteins, the structures are virtually identical. VPS29 in human and mouse is composed of a mixed α/β structure and consists of a central β-sandwich of 2 β-sheets, one surrounded by 2 α-helices and a single α-helix extending from the other β-sheet (Collins et al. 2005; Wang et al. 2005). Analysis of both mouse and human VPS29 crystal structures determined that they have a structural resemblance to phosphoesterases, in particular to a family of divalent, metal-containing phosphoesterases (DMPs), including serine/threonine PPP phosphatases, all of which contain a conserved active site that binds divalent metal ions (Collins et al. 2005). The active site of mouse and human VPS29 closely resembles a phosphodiesterase from Archaea called MJ0936 (Collins et al. 2005; Wang

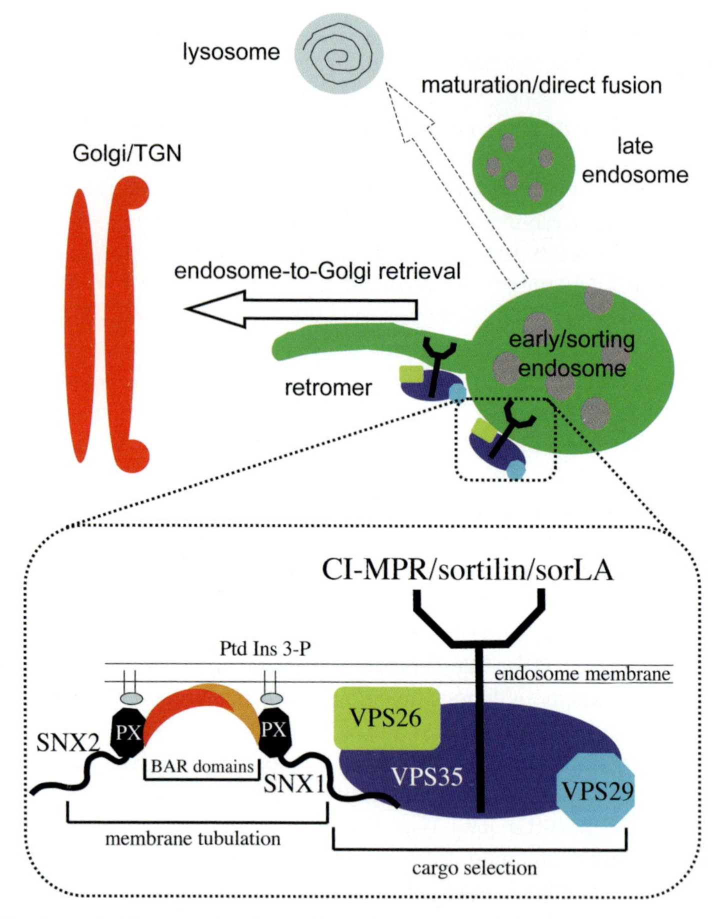

Fig. 1 A schematic diagram of endosomal protein sorting and the retromer complex. In mammalian cells, retromer is recruited to the endosome partly through the action of the PX domains in the sorting nexin proteins SNX1 and SNX2, which drive membrane tubulation through their BAR domains and self-assembly activity. VPS35/29/26 interact to sort cargo proteins such as the CI-MPR, sortilin and SorLA, which traffic between endosomes and the Golgi/trans-Golgi network (TGN)

et al. 2005). The PPP phosphatases have a high degree of conservation around the metal-binding site, with both the metal-binding residues and the phosphate-binding residues conserved (Barton et al. 1994).

The active site of mouse VPS29 was crystallised bound to two Mn2+ ions through conserved histidine, asparagine and aspartate residues like other DMPs, but in mouse VPS29, the metal bridging residues for each Mn2+ ion coordinated were found to be reversed compared to known DMPs. Both mouse and human VPS29 could coordinate metal ions in their active site pockets (Collins et al. 2005; Wang et al. 2005). The conserved exterior hydrophobic patches present on human VPS29 may be integral to VPS29 interacting with other retromer components (Wang et al.

2005), and when these residues were mutated in either yeast Vps29p (V109) or mammalian VPS29 (V90), the resulting mutants were unable to bind Vps35p or VPS35, respectively (Collins et al. 2005).

The crystal structure of human VPS26 has been recently solved and was demonstrated to bear a structural resemblance to a group of proteins called arrestins (Shi et al. 2006; Collins et al. 2008). Equivalent to known arrestin structures, hVPS26 is composed of two β-sandwich domains, both of which are heavily curved and have a polar core in the middle. Human VPS26 also contains a conserved basic patch and two acidic patches. Mutations were made in hVPS26 to identify the binding site for hVPS35, and mutation of I235 and M236 or loss of amino acid residues 238–246 (forming a loop on the far carboxyl-terminal end of the hVPS26 structure) caused a loss of interaction with VPS35 in the yeast 2-hybrid system. The relevance of the loop section was confirmed by immunoprecipitation studies, where tagged mutant VPS26 missing the loop residues or I235/M236 could not co-precipitate hVPS35 like wild-type VPS26 (Shi et al. 2006). Mutation of the equivalent yeast Vps26p residues caused a defect in the ability of these mutants to complement the CPY sorting defect seen in *vps26Δ* yeast. These data therefore suggest the loop section on the carboxyl-terminus is crucial for VPS26 function as part of retromer in both yeast and humans.

5 Membrane Association of Retromer

Retromer is peripherally associated with endosomal membranes. The mechanisms that underlie this association are not yet fully understood. The PX domains in Vps5p/Vps17p and their mammalian counterparts, SNX1/SNX2, are believed to be important in promoting membrane recruitment or targeting. Experiments in yeast have shown that the phosphotidyl inositol 3-kinase (Ptd Ins-3 kinase) enzyme encoded by the *VPS34* gene plays an important role in targeting/recruitment of Vps5p/Vps17p to the endosomal membrane (Burda et al. 2002). The activity of the Vps34 protein is regulated by the Vps30p/Vps38p complex, which binds to Vps34p, stimulating production of Ptd Ins-3P and thereby facilitating recruitment of Vps5p/Vps17p (Kihara et al. 2001; Burda et al. 2002). Similarly, in mammalian cells, treatment with the drug wortmannin, which inhibits mammalian PI-3 kinases, results in redistribution of SNX1 and SNX2 to the cytoplasm, indicating a requirement for Ptd Ins-3P in regulating membrane targeting/recruitment of SNX1 and SNX2 (Kurten et al. 2001; Cozier et al. 2002).

A curious aspect of the membrane recruitment of Vps5p/Vps17p is that, in yeast, Vps30p has a dual function, being required for not only vacuole protein sorting but also autophagy (Kametaka et al. 1998). In autophagy, Vps30p also regulates the activity of Vps34p, but this time Vps30p is complexed with a different protein, namely Atg14p (Kihara et al. 2001). Vps30p is conserved and the mammalian homologue, Beclin-1, has been shown to function in autophagy in mammalian cells. Beclin-1 can rescue the autophagy defect in *vps30Δ* yeast but cannot rescue the vacuole protein-sorting defect (Liang et al. 1999) suggesting that the role of Vps30p in autophagy is conserved in evolution but its role in vacuole protein sorting is not. It is noteworthy that Vps38p is not conserved, with no clear homologues present in higher eukaryotes such as mammals. It is possible, therefore, that the regulatory role of the Vps30p/Vps38p complex is unique to *S. cerevisiae*.

How the Vps35p/29p/26p subcomplex is recruited to the membrane is less clear. These proteins do not possess PX domains or any other obvious lipid-binding domains. In yeast, deletion of *VPS35* causes Vps29p to become cytoplasmic, and deletion of *VPS26* results in Vps35p becoming partly soluble, suggesting that the interaction between Vps35p and Vps26p can facilitate the membrane association of the Vps35p/Vps29p/Vps26p complex (Seaman et al. 1998; Reddy and Seaman 2001). As Vps26p can remain on the membrane after deletion of *VPS35* and/or *VPS29*, it seems likely that Vps26p can interact directly with the endosomal membrane, although the molecular mechanisms that govern this interaction have yet to be determined.

6 Retromer Functions in Endosome-to-Golgi Retrieval

The studies conducted on retromer mutants in yeast strongly suggested that retromer functions in endosome-to-Golgi retrieval, being required for the proper recycling of the Vps10 protein. However, there is no direct assay for endosome-to-Golgi retrieval in yeast, only assays that can measure the consequences of a lack of retrieval. For example, the CPY sorting defect in a retromer mutant is believed to be the result of a lack of Vps10p in the late-Golgi. Studies conducted in mammalian cells have more directly addressed the issue of the role of retromer in endosome-to-Golgi transport (Seaman 2004; Arighi et al. 2004).

Firstly, retromer has been localised to endosomes in mammalian cells. Significant colocalisation of hVPS26 with the early endosomal markers, rab5 and EEA1, was observed by immunofluorescence (Seaman 2004). Using the greater resolving power of immuno-gold EM, both hVPS26 and SNX1 were found on multivesicular body endosomes. In some instances, the structures labelled with anti-retromer antibodies were found to be tubular (Seaman 2004; Arighi et al. 2004). Recent detailed EM studies have shown that retromer is localised to "exit sites" at early endosomes (Popoff et al. 2007). Mouse cells derived from a transgenic VPS26 (Hβ58) – /– knock-out have swollen lysosomes and accumulate unesterified cholesterol, which is similar to the defect in cells lacking the function of the Niemann-Pick protein, NPC1 (see Fig. 2).

Analysis of the VPS26 (Hβ58) – /– cells and studies conducted using small interfering (si) RNA to "knockdown" expression of hVPS26 demonstrated that loss of retromer function results in mislocalisation of the CI-MPR to either lysosomes or the plasma membrane (Seaman 2004; Arighi et al. 2004). Failure to retrieve the CI-MPR from endosomes would be predicted to result in either degradation in lysosomes or increased localisation to the cell surface, so the observations made

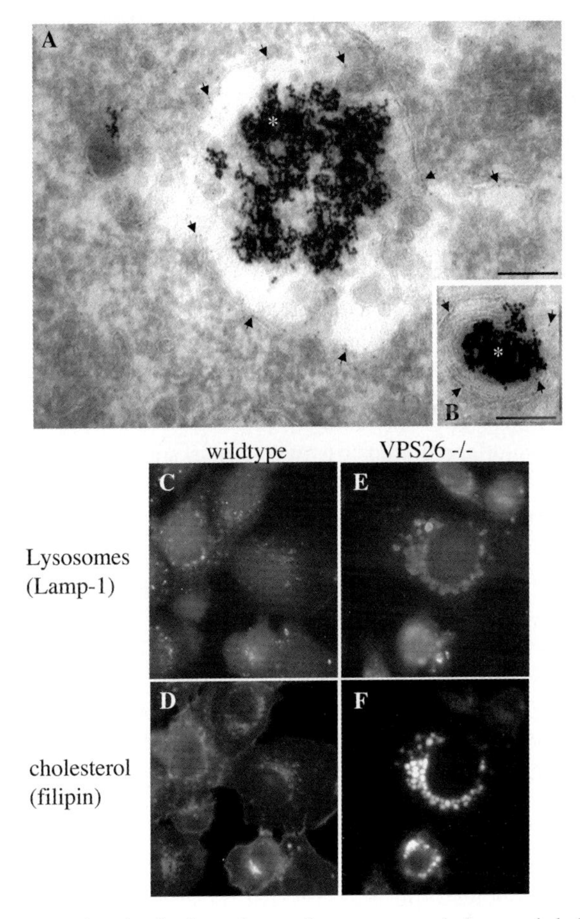

Fig. 2 Loss of retromer function leads to aberrant lysosome morphology and cholesterol accumulation. **A** and **B**. Electron micrographs of BSA-gold- (indicated with *) loaded lysosomes labeled with antibodies against Lamp-1 (arrows). Cells were incubated with BSA-gold for four hours and chased overnight before fixation and processing for cryo-electron microscopy. In **A**, cells from a transgenic VPS26 −/− mouse were used whereas in **B** wildtype cells were analysed. The scale bar is 200 nm in both micrographs. Cells lacking VPS26 display massively enlarged lysosomes. In **C** and **E**, lysosomes labeled with Lamp-1 antibodies are shown in wildtype and VPS26 −/− cells, respectively; in **D** and **F**, cholesterol is visualized using filipin. Loss of retromer function due to knockout of VPS26 results in an accumulation of cholesterol in lysosomes, a phenotype observed in some lysosomal storage disorders

in VPS26 (Hβ58) −/− cells and after siRNA knockdown of hVPS26 are consistent with retromer functioning in endosome-to-Golgi transport. Loss of retromer function also results in a failure to properly mature the lysosomal hydrolase, cathepsin D (Seaman 2004; Arighi et al. 2004), mirroring the results obtained in yeast, where retromer mutants display strong CPY sorting defects. Studies on SNX1 revealed that it is required for the endosome-to-Golgi retrieval of the CI-MPR and acts in concert with two other sorting nexins, SNX5 and SNX6 (Carlton et al. 2004; Wassmer et al. 2007).

Using reporter constructs based upon the T-cell marker, CD8, the trafficking of two chimeras was studied in control and retromer knockdown cells. The two chimeras were CD8-CI-MPR and CD8-furin. These reporters had identical lumenal and transmembrane domains but differed in their respective cytoplasmic domains, having the cytoplasmic tails of the CI-MPR and furin, respectively. As both these proteins cycle between the Golgi and endosomes, they are obvious candidates for trafficking in a retromer-mediated pathway. Using an antibody uptake assay, the trafficking of the CD8-CI-MPR and CD8-furin reporters was examined in both control and hVPS26 knockdown cells. Loss of hVPS26 expression resulted in a block in endsome-to-Golgi transport of the CD8-CI-MPR, consistent with the findings for the endogenous CI-MPR. Interestingly, the trafficking of the CD8-furin reporter did not appear to be affected after loss of retromer, suggesting that retromer plays a cargo-selective role in endosome-to-Golgi trafficking (Seaman 2004). Indeed, a direct interaction between the cytoplasmic tail of the CI-MPR and hVPS35 was detected using the two-hybrid system, confirming that retromer plays a crucial role in the endosome-to-Golgi retrieval of the CI-MPR (Arighi et al. 2004).

Exploiting the CD8-reporter protein system to investigate the retrieval of the CD8-CI-MPR reporter, a conserved hydrophobic motif (Trp-Leu-Met) present in the CI-MPR tail was shown to be necessary for its endosome-to-Golgi retrieval and for in vivo association with retromer, as determined by native immunoprecipitation (Seaman 2007). As the CI-MPR is present in Golgi membranes, endosomes and the plasma membrane, the binding of hVPS35 to the CI-MPR tail must somehow be regulated. How this is achieved is currently unknown.

Using the CD8-reporter system, it was observed that a CD8 chimera carrying the tail of the transmembrane protein, sortilin, is also dependent upon retromer for its endosome-to-Golgi trafficking (Seaman 2004), and similar observations have been made for the native sortilin protein (Canuel et al. 2008). Sortilin is the closest mammalian homologue to Vps10p and, therefore, might be expected to traffic in a retromer-dependent fashion. These data conclusively demonstrate that retromer is required for the endosome-to-Golgi transport of the CI-MPR and other proteins such as sortilin. Interestingly, the cytoplasmic tail of sortilin contains a motif, Phe-leu-Val, which is similar to the Trp-leu-Met motif in the CI-MPR and which is also necessary for the retrieval of a CD8-sortilin reporter (Seaman 2007).

More recently, retromer has been shown to mediate the trafficking of SorLA, another member of the sortilin family of type I membrane proteins (Nielsen et al. 2007). SorLA (which is also known as SORL1) binds to the amyloid precursor protein (APP) and is required to prevent cleavage of APP to the neurotoxic pro-aggregatory Aβ peptide that is responsible for Alzheimer's disease (Andersen et al. 2005). Retromer can co-immunoprecipitate with a CD8-SorLA reporter protein, and loss of retromer function by siRNA knockdown of VPS26 results in a defect in the endosome-to-Golgi retrieval of CD8-SorLA (see Fig. 3). The role of retromer in mediating the trafficking of SorLA has, therefore, firmly established retromer and endosome-to-Golgi retrieval as an important process in the development of

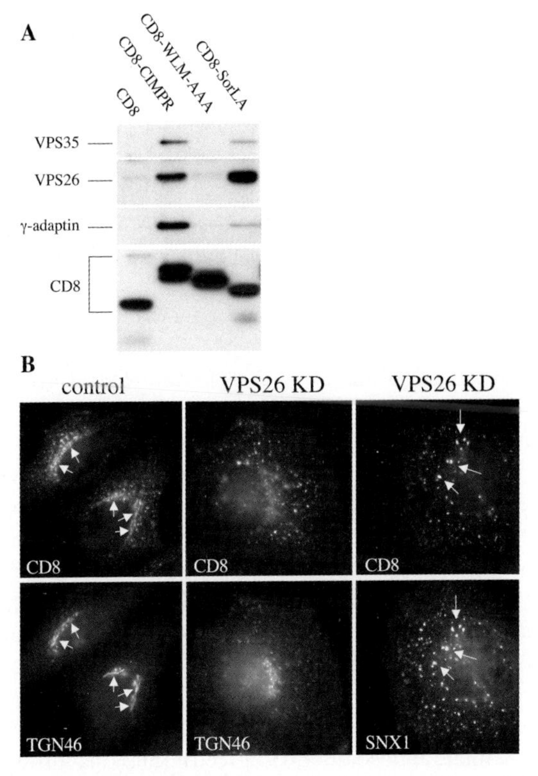

Fig. 3 Retromer is required for proper trafficking of the SorLA protein. **A**. Cells expressing CD8-reporter proteins were lysed and the CD8 reporter was immunoprecipitated using anti-CD8. The retromer proteins VPS35 and VPS26 co-immunoprecipitate with CD8 reporters carrying the tail of the CI-MPR or SorLA but not with native CD8 or a mutant of the CI-MPR tail (WLM-AAA) that does not bind retromer and does not retrieve from endosomes to the Golgi (Seaman 2007). **B**. Cells expressing CD8-SorLA were subjected to siRNA knockdown (KD) of VPS26 and an antibody uptake assay was used to evaluate endosome-to-Golgi retrieval of the CD8-SorLA protein. In control cells, the CD8-SorLA construct can traffic from endosomes to the Golgi (shown by TGN46 protein) but, after loss of VPS26 expression, the CD8-SorLA protein accumulates within endosomes that are positive for the SNX1 protein. Therefore, efficient retrieval of the CD8-SorLA protein requires retromer

Alzheimer's disease. A reduction in retromer expression has been demonstrated in neuronal tissue from Alzheimer's patients, further establishing retromer as a key player in the pathogenesis of Alzheimer's (Small et al. 2005).

Exactly how retromer mediates the retrieval of proteins from endosomes to the Golgi is currently unknown. In many respects, retromer possesses the defining characteristics of a vesicle coat. Retromer can bind to cargo (e.g., the CI-MPR) and can also self-assemble through the action of the Vps5 protein (in yeast) or SNX1 and associated sorting nexins (in mammalian cells). To date, however, there is no conclusive evidence that retromer mediates the budding of vesicles in a similar way to clathrin coats. One possibility is that retromer-mediated endosome-to-Golgi retrieval occurs via tubules rather than discrete vesicles. The SNX1 and SNX2 proteins have been shown to associate with thread-like tubules that emanate from endosomes. Using green fluorescent protein (GFP)-tagged SNX1, these tubules were found to be highly dynamic and were observed "growing" from endosomes (Carlton et al. 2004; Arighi et al. 2004). SNX1 and SNX2 have Bin/Amphiphysin/ Rvs (BAR) domains in their carboxyl-terminal halves (Peter et al. 2004). BAR domains have the ability to cause tubulation of liposomes in vitro, raising the prospect that SNX1 with SNX2 could be promoting tubulation of endosomal membranes through the combined action of self-assembly and the BAR domain. The observation that the tubules are often several microns long and very straight hints at a requirement for the cytoskeleton in maintaining/stabilising the tubules. It has been demonstrated that endosome-to-Golgi retrieval of the CI-MPR is facilitated by microtubules (MTs; Itin et al. 1999), but it is currently unknown if retromer has any direct interaction with MTs or MT-associated proteins.

Contributing to retromer function in endosome-to-Golgi retrieval are accessory proteins that interact with retromer. So far, two such accessory proteins have been identified. Firstly, in yeast, Grd19p is a member of the sorting nexin family and is homologous to SNX3 in mammals. The Ftr1p protein that is required for iron homeostasis in yeast is recognised by Grd19p, which also interacts with retromer to direct Ftr1p into the retromer-mediated endosome-to-Golgi pathway (Strochlic et al. 2007). Grd19p therefore functions as a cargo-specific adaptor, expanding the repertoire of retromer in yeast beyond Vps10p.

In mammalian cells, the second retromer accessory protein, EHD1, interacts with retromer to facilitate endosome-to-Golgi retrieval by stabilising SNX1-positive tubules. EHD1 is upregulated in cells derived from the mouse VPS26 (Hβ58) −/− knockout cells and partially colocalises with retromer on endosomes. Loss of EHD1 expression by siRNA inhibition causes a kinetic delay in endosome-to-Golgi retrieval and reduces the number of SNX1-positive tubules by ~60%, whereas over-expression of dominant negative EHD1 mutants causes retromer to redistribute from endosomes and potently blocks endosome-to-Golgi retrieval (Gokool et al. 2007b). There are likely to be further discoveries of retromer-interacting accessory proteins in the near future, some of which may play a direct role in neurodegenerative disease.

7 The Unanswered Questions

Studies on retromer are currently ongoing and there are many questions that remain to be answered. How retromer and its function in endosome-to-Golgi traffic are regulated is currently unknown. Regulation of membrane recruitment could provide one level of regulation. There is a requirement for Ptd Ins-3P in the targeting/ recruitment of the Vps5p/Vps17p, SNX1/SNX2 component of retromer. Therefore, the regulated production of Ptd Ins-3P by a Ptd Ins 3-kinase is one potential avenue of regulation. There may be a requirement for phosphorylation in regulating retromer activity. Both Vps5p and Vps17p are phosphoproteins (Horazdovsky et al. 1997; Kohrer and Emr 1993); however, what role the phosphorylation of Vps5p/Vps17p plays in regulating retromer function is not presently known.

Additionally it is unknown whether retromer is functioning in the same pathway as other proteins that have been previously shown to function in endosome-to-Golgi retrieval. The small GTPase rab9 has a well-established role in the retrieval of the CI-MPR (Riederer et al. 1994). Currently, however, it is not known whether retromer and rab9 function together. No clear colocalisation was observed between retromer and rab9 (Seaman 2004); therefore, if they do function in the same pathway, it seems likely that they may be acting sequentially. Retromer acts to select cargo into the vesicles/tubules for retrieval; rab9 would then act to regulate docking and fusion of the vesicle/tubule with the TGN.

In summary, retromer has been characterised both in yeast and mammalian cells and has been shown to play a vital role in the endosome-to-Golgi retrieval of vacuole/lysosome hydrolase receptors. Other proteins that traffic in the post-Golgi endocytic system, such as SorLA, are also recognised by retromer, placing retromer at the heart of the cellular machinery that regulates APP localisation and processing, with clear implications for the pathogenesis of Alzheimer's disease. Retromer is still a relatively new "coat" complex and there remains much to learn about how retromer performs its task. With studies ongoing in many labs, it will be interesting to see how this story develops and what new results are published in the coming years. It seems likely that additional accessory proteins that interact with retromer will be identified, and these may provide further insights into the role of retromer in endosome-to-Golgi retrieval and neurodegenerative disease.

Acknowledgements Matthew Seaman is the recipient of a Medical Research Council Senior Research Fellowship. Electron microscopy was performed by Dr. Margaret Lindsay.

References

Andersen OM, Reiche J, Schmidt V, Gotthardt M, Spoelgen R, Behlke J, von Arnim CA, Breiderhoff T, Jansen P, Wu X, Bales KR, Cappai R, Masters CL, Gliemann J, Mufson EJ, Hyman BT, Paul SM, Nykjaer A, Willnow TE (2005) Neuronal sorting protein-related receptor sorLA/LR11 regulates processing of the amyloid precursor protein. Proc Natl Acad Sci USA. 102:13461–13466

Arighi CN, Hartnell LM, Aguilar RC, Haft CR, Bonifacino JS (2004) Role of the mammalian retromer in sorting of the cation-independent mannose 6-phosphate receptor. J Cell Biol 165:123–133

Barton GJ, Cohen PT, Barford D (1994) Conservation analysis and structure prediction of the protein serine/threonine phosphatases. Sequence similarity with diadenosine tetraphosphatase from Escherichia coli suggests homology to the protein phosphatases. Eur J Biochem 220:225–237

Bryant NJ, Stevens TH (1998) Vacuole biogenesis in Saccharomyces cerevisiae: protein transport pathways to the vacuole. Microbiol Mol Biol Rev 62:230–247

Burda P, Padilla SM, Sarkar S, Emr SD (2002) Retromer function in endosome-to-Golgi retrograde transport is regulated by the yeast Vps34 Ptd Ins 3-kinase. J Cell Sci 115:3889–900

Canuel M, Lefrancois S, Zeng J, Morales CR (2008) AP-1 and retromer play opposite roles in the trafficking of sortilin between the Golgi apparatus and the lysosomes. Biochem Biophys Res Commun 366:724–730

Carlton J, Bujny M, Peter BJ, Oorschot VM, Rutherford A, Mellor H, Klumperman J, McMahon HT, Cullen PJ (2004) Sorting nexin-1 mediates tubular endosome-to-TGN transport through coincidence sensing of high- curvature membranes and 3-phosphoinositides. Curr Biol 14:1791–1800

Cereghino J.-L., Marcusson EG, Emr SD (1995) The cytoplasmic tail domain of the vacuolar sorting receptor Vps10p and a subset of *VPS* gene products regulate receptor stability, function, and localization. Mol Biol Cell 6:1089–1102

Collins BM, Skinner CF, Watson PJ, Seaman MN, Owen DJ (2005) Vps29 has a phosphoesterase fold that acts as a protein interaction scaffold for retromer assembly. Nat Struct Mol Biol 12:594–602

Collins BM, Norwood SJ, Kerr MC, Mahony D, Seaman MN, Teasdale RD, Owen DJ (2008) Structure of Vps26B and mapping of its interaction with the retromer protein complex. Traffic 9:366–379

Cooper AA, Stevens TH (1996) Vps10p cycles between the late-Golgi and prevacuolar compartments in its function as the sorting receptor for multiple yeast vacuolar hydrolases. J Cell Biol 133:529–542

Costaguta G, Stefan CJ, Bensen ES, Emr SD, Payne GS (2001) Yeast GGA coat proteins function with clathrin in Golgi to endosome transport. Mol Biol Cell 12:1885–1896

Cozier GE, Carlton J, McGregor AH, Gleeson PA, Teasdale RD, Mellor H, Cullen PJ (2002) The Phox homology (PX) domain-dependent, 3-phosphoinositide-mediated association of Sorting Nexin-1 with an early sorting endosomal comparment is required for its ability to regulate epidermal growth factor receptor degradation. J Biol Chem 277:48730–48736

Dell'Angelica EC, Puertollano R, Mullins C, Aguilar RC, Vargas JD, Hartnell LM, Bonifacino JS (2000) GGAs: a family of ADP ribosylation factor-binding proteins related to adaptors and associated with the Golgi complex. J Cell Biol 149:81–94

Gokool S, Tattersall D, Reddy JV, Seaman MN (2007a) Identification of a conserved motif required for Vps35p/Vps26p interaction and assembly of the retromer complex. Biochem J 408:287–295

Gokool S, Tattersall D, Seaman MN (2007b) EHD1 interacts with retromer to stabilize SNX1 tubules and facilitate endosome-to-Golgi retrieval. Traffic 8:1873–1886

Horazdovsky BF, Davies BA, Seaman, MNJ, McLaughlin SA, Yoon S.-H., Emr SD (1997) A sorting nexin-1 homologue, Vps5p, forms a complex with Vps17p and is required for recycling the vacuolar protein-sorting receptor. Mol Biol Cell 8:1529–1541

Hwang S, Benjamin LE, Oh B, Rothstein JL, Ackerman SL, Beddington RS, Solter D, Knowles BB (1996) Genetic mapping and embryonic expression of a novel maternally transcribed gene, Mem3. Mamm Genome 7:586–590

Itin C, Ulitzur N, Muhlbauer B, Pfeffer SR (1999) Mapmodulin, cytoplasmic dynein, and microtubules enhance the transport of mannose 6-phosphate receptors from endosomes to the trans-Golgi network. Mol Biol Cell 10:2191–2197

Kametaka S, Okano T, Ohsumi M, Ohsumi Y (1998) Apg14p and Apg6/Vps30p form a protein complex essential for autophagy in the yeast, Saccharomyces cerevisiae. J Biol Chem 273:22284–22291

Kihara A, Noda T, Ishihara N, Ohsumi Y (2001) Two distinct Vps34 phosphotidylinositol 3-kinase complexes function in autophagy and carboxypeptidase Y sorting in Saccharomyces cerevisiae. J Cell Biol 152:519–530

Kohrer K, Emr SD (1993) The yeast VPS17 gene encodes a membrane-associated protein required for the sorting of soluble vacuolar hydrolases. J Biol Chem 268:559–569

Kornfeld S (1992) Structure and function of the mannose 6-phosphate/insulin-like growth factor II receptors. Annu Rev Biochem 61:307–330

Kurten RC, Cadena DL, Gill GN (1996) Enhanced degradation of EGF receptors by a sorting nexin, SNX1. Science 272:1008–1010

Kurten RC, Eddington AD, Chowdhury P, Smith RD, Davidson AD, Shank BB (2001) Self-assembly and binding of a sorting nexin to sorting endosomes. J Cell Sci 114:1743–1756

Lee JJ, Radice G, Perkins C, Costantini F (1992) Identification and characterization of a novel, evolutionary conserved gene disrupted by the murine Hβ58 embryonic lethal transgene insertion. Development 115:277–288

Liang XH, Jackson S, Seaman M, Brown K, Kempkes B, Hibshoosh H, Levine B (1999) Induction of autophagy and inhibition of tumourigenesis by beclin 1. Nature 402:672–676

Marcusson EG, Horazdovsky BF, Cereghino J.-L., Gharakhanian E, Emr SD (1994) The sorting receptor for yeast vacuolar carboxypeptidase Y is encoded by the VPS10 gene. Cell 77:579–586

Misra S, Puertollano R, Kato Y, Bonifacino JS, Hurley JH (2002) Structural basis for acidic-cluster-dileucine sorting-signal recognition by VHS domains. Nature 21:933–937

Ni X, Canuel M, Morales CR (2006) The sorting and trafficking of lysosomal proteins. Histol Histopathol 21:899–913

Nielsen MS, Gustafsen C, Madsen P, Nyengaard JR, Hermey G, Bakke O, Mari M, Schu P, Pohlmann R, Dennes A, Petersen CM (2007) Sorting by the cytoplasmic domain of the amyloid precursor protein binding receptor SorLA. Mol Cell Biol 27:6842–6851

Nothwehr SF, Hindes AH (1997) The yeast VPS5/GRD2 gene encodes a sorting nexin-1-like protein required for localizing membrane proteins to the late Golgi. J Cell Sci 110:1063–1072

Nothwehr SF, Bruinsma P, Strawn LS (1999) Distinct Domains within Vps35p mediate retrieval of two different cargo proteins from the yeast prevacuolar/endosomal compartment. Mol Biol Cell 10:875–890

Nothwehr SF, Ha S.-A., Bruinsma P (2000) Sorting of Yeast Membrane Proteins into an Endosomal-to-Golgi Pathway Involves Direct Interaction of their Cytosolic Domains with Vps35p. J Cell Biol 151:297–309

Peter BJ, Kent HM, Mills IG, Vallis Y, Butler PJ, Evans PR, McMahon HT (2004) BAR domains as sensors of membrane curvature: The Amphiphysin BAR structure. Science 303:495–499

Popoff V, Mardones GA, Tenza D, Rojas R, Lamaze C, Bonifacino JS, Raposo G, Johannes L (2007) The retromer complex and clathrin define an early endosomal retrograde exit site. J Cell Sci 120:2022–2031

Reddy JV, Seaman MNJ (2001) Vps26p, a component of retromer, directs the interactions of Vps35p in Endosome-to-Golgi retrieval. Mol Biol Cell 12:3242–3256

Renfrew-Haft C, Sierra M, Bafford R, Lesniak MA, Barr VA, Taylor SI (2000) Human Orthologs of Yeast Vacuolar Protein Sorting Proteins Vps26, 29 and 35: Assembly into Multimeric complexes. Mol Biol Cell 11:4105–4116

Restrepo R, Zhao X, Peter H, Zhang BY, Arvan P, Nothwehr SF (2007) Structural features of vps35p involved in interaction with other subunits of the retromer complex. Traffic 8:1841–1853

Riederer MA, Soldati T, Shapiro AD, Lin J, Pfeffer SR (1994) Lysosome biogenesis requires Rab9 function and receptor recycling from endosome to the trans-Golgi network. J Cell Biol 125:573–582

Seaman MN (2007) Identification of a novel conserved sorting motif required for retromer-mediated endosome-to-TGN retrieval. J Cell Sci 120:2378–2389

Seaman, MNJ (2004) Cargo-selective endosomal sorting for retrieval to the Golgi requires retromer. J Cell Biol 165:111–122

Seaman MNJ, Williams HP (2002) Identification of the functional domains of yeast sorting nexins Vps5p and Vps17p. Mol Biol Cell 13:2826–2840

Seaman MNJ, Marcusson EG, Cereghino J-L, Emr SD (1997) Endosome to Golgi retrieval of the vacuolar protein sorting receptor, Vps10p, requires the function of VPS29, VPS30 and VPS35 gene products. J Cell Biol 137:79–92

Seaman MNJ, McCaffery JM, Emr SD (1998) A Membrane coat Complex Essential for Endosome-to-Golgi retrograde transport in Yeast. J Cell Biol 141:665–681

Shi H, Rojas R, Bonifacino JS, Hurley JH (2006) The retromer subunit Vps26 has an arrestin fold and binds Vps35 through its C-terminal domain. Nature Struct Mol Biol 13:540–548

Shiba T, Takatsu H, Nogi T, Matsugaki N, Kawasaki M, Igarashi N, Suzuki M, Kato R, Earnest T, Nakayama K, Wakatsuki S (2002) Structural basis for the recognition of acidic cluster dileucine sequences by GGA1. Nature 21:937–941

Small SA, Kent K, Pierce A, Leung C, Kang MS, Okada H, Honig L, Vonsattel JP, Kim TW (2005) Model-guided microarray implicates the retromer complex in Alzheimer's disease. Ann Neurol 58:909–919

Strochlic TI, Setty TG, Sitaram A, Burd CG (2007) Grd19/Snx3p functions as a cargo-specific adapter for retromer-dependent endocytic recycling. J Cell Biol 177:115–125

Sturley SL, Patterson MC, Balch W, Liscum L (2004) The pathophysiology and mechanisms of NP-C disease. Biochim Biophys Acta 1685:83–87

Wang D, Guo M, Liang Z, Fan J, Zhu Z, Zang J, Zhu Z, Li X, Teng M, Niu L, Dong Y, Liu P (2005) Crystal structure of human vacuolar protein sorting protein 29 reveals a phosphodiesterase/nuclease-like fold and two protein-protein interaction sites. J Biol Chem 280:22962–22967

Wassmer T, Attar N, Bujny MV, Oakley J, Traer CJ, Cullen PJ (2007) A loss-of-function screen reveals SNX5 and SNX6 as potential components of the mammalian retromer. J Cell Sci 120:45–54

Westphal V, Marcusson EG, Winther JR, Emr SD, van den Hazel HB (1996) Multiple pathways for vacuolar sorting of yeast proteinase A. J Biol Chem 271:11865–11870

Yu JW, Lemmon MA (2001) All Phox homology (PX) domains from Saccharomyces cerevisiae specifically recognize phosphotidylinositol 3-phosphate. J Biol Chem 276:44179–44184

Zhao X, Nothwehr S, Lara-Lemus R, Zhang BY, Peter H, Arvan P (2007) Dominant-negative behavior of mammalian Vps35 in yeast requires a conserved PRLYL motif involved in retromer assembly. Traffic 8:1829–1840

Regulation of Endocytic Trafficking of Receptors and Transporters by Ubiquitination: Possible Role in Neurodegenerative Disease

Alexander Sorkin

Abstract Ubiquitination has recently emerged as the major regulatory mechanism of endocytic trafficking of transmembrane proteins. Ubiquitin-controlled trafficking and endocytosis regulate the function of various receptors, channels and transporters in neurons, and deregulation of the ubiquitination system is associated with neurodegenerative diseases. Hence, we will focus on recent advances in understanding the mechanisms and functional roles of ubiquitination of two families of transmembrane proteins: (1) receptor tyrosine kinases, using the receptor for epidermal growth factor (EGFR) as a prototypic member of the family; and (2) monoamine transporters, using an example of the plasma membrane dopamine transporter (DAT). Both these families of receptors and transporters are intimately involved in brain development, regulation of survival signaling in adult neurons, neurotransmission, neuronal cytotoxicity and neurodegeneration. Endocytosis regulates the duration and intensity of the EGFR signaling. Endocytosis of DAT controls the re-uptake of dopamine in dopaminergic neurons, thus regulating dopamine neurotransmission in the brain. Our recent studies revealed unexpected similarities in the regulation of endocytosis of these two structurally distinct families of proteins by ubiquitination. We have mapped ubiquitin conjugation sites in the EGFR and demonstrated that mutation of these sites results in inhibition of the lysosomal targeting and degradation of EGFR. However, EGFR ubiquitination appears not to be essential for the internalization step of the EGFR trafficking. Surprisingly, we have recently found that DAT is also ubiquitinated and the extent of its ubiquitination is dramatically increased upon activation of protein kinase C (PKC). The ubiquitination sites in DAT were also mapped by mass spectrometry. Mutations of a cluster of three lysines in the N-terminal tail of DAT blocked the clathrin-mediated endocytosis of DAT. Screening of the library of small interfering RNAs revealed that NEDD4-2 is an E3 ubiquitin ligase responsible for ubiquitination of DAT and necessary for PKC-dependent DAT endocytosis. Thus, our studies revealed that both EGFR and DAT are ubiquitinated

A. Sorkin
Department of Pharmacology, University of Colorado Denver Health Sciences Center, Aurora Colorado 80045, USA

P. St. George-Hyslop et al. (eds.) *Intracellular Traffic and Neurodegenerative Disorders*, Research and Perspectives in Alzheimer's Disease,

at the plasma membranes and endosomes, and this ubiquitination regulates their turnover and subcellular localization. Interestingly, both EGFR and DAT are modified by Lys63-linked poly-ubiquitin chains. We hypothesize that short, Lys63-linked chains are the major ubiquitin-based molecular signals operating during endocytic trafficking in mammalian cells.

1 Introduction

The activities of neuronal cells and their survival are controlled by various receptor, channel and transporter proteins present at the surface of these neurons, where they interact with their ligands and substrates. Various classes of transport proteins essential for synaptic transmission and neuronal signaling function in the intracellular compartments, such as synaptic vesicles and endosomes. For example, receptor tyrosine kinases (RTKs), such as TrkA receptors for the nerve growth factor, require endocytosis at the distal axonal processes and an axonal transport of TrkA signaling complexes in endosomes for the retrograde survival signaling in the neuronal soma (Zweifel et al. 2005). Endocytosis of APP appears to be necessary for the neuronal activity-dependent extracellular accumulation of the amyloid-β peptide (Cirrito et al. 2008). Thus, aberrant endocytic trafficking leading to mis-localization of transmembrane proteins within the neuronal cell often underlies the mechanisms responsible for the development of the neurodegenerative disease.

Rapid and dynamic regulation of the amounts of receptors and transport proteins at the plasma membrane and intracellular membrane compartments in the synapse and extrasynaptically is achieved by means of selective endocytosis and recycling of these proteins. Many receptors and transport proteins are rapidly endocytosed in a constitutive or stimuli-dependent manner. Subsequently, the internalized transmembrane proteins (i.e., cargo) are either recycled back from endosomes to the plasma membrane, or accumulate in specialized compartments, such as synaptic vesicles and endosomes, or are sorted to lysosomes for degradation. The mechanisms of endocytosis and post-endocytic trafficking of membrane proteins have been extensively studied over the last 30 years; however, molecular details of many steps of these processes remain poorly understood.

Posttranslational modification of transmembrane proteins by the covalent attachment of ubiquitin has recently emerged as the major regulatory mechanism of endocytic trafficking of these proteins. Many of the original observations of ubiquitination of the endocytic cargo and regulation of endocytosis by ubiquitination were made in yeast (Hicke and Riezman 1996; Kolling and Hollenberg 1994). Among mammalian ubiquitinated cargo are RTKs; Notch and its transmembrane ligands, cytokine and interferon receptors; various channels and transporters; G protein coupled receptors (GPCR); and other types of transmembrane proteins (Hicke and Dunn 2003; Staub and Rotin 2006). Our laboratory is focusing on the mechanisms and functional roles of ubiquitination of two classes of molecules: (1) RTKs, using a prototypic member of the family, the epidermal growth factor (EGF) receptor (EGFR) as an experimental model; and (2) plasma membrane solute transporters, using the plasma membrane dopamine transporter (DAT) as an experimental model.

2 Modification of Proteins by Ubiquitin

Ubiquitination is a posttranslational modification that mediates the covalent conjugation of ubiquitin, a highly conserved protein of 76 amino acids, to protein substrates. Ubiquitination was originally thought to target proteins for degradation by the 26*S* proteasome (Hershko and Ciechanover 1992). However, the role of ubiquitination in many non-proteosomal processes in the cell, including membrane trafficking, DNA repair, and transcription, has been recently revealed (Mukhopadhyay and Riezman 2007; Pickart and Fushman 2004). The observations of an abnormal enrichment of inclusion bodies with ubiquitin in Huntington's disease and many other neurodegenerative disorders, including Alzheimer's and Parkinson's diseases (Lowe et al. 1988; Mayer et al. 1989), have suggested that dysfunction in ubiquitin metabolism may contribute to the pathogenesis of these diseases (DiFiglia et al. 1997; Ross and Pickart 2004).

The mechanism of ubiquitination involves the sequential action of several enzymes. In the initial step, the E1 ubiquitin-activating enzyme forms a thioester bond between its catalytic cysteine and the carboxyl group of Gly76 of ubiquitin in an ATP-dependent manner. The ubiquitin molecule is then transferred to an E2 ubiquitin-conjugating enzyme, which also forms a thioester bond between its cysteine and ubiquitin. Finally, ubiquitin is transferred to a lysine residue of the substrate with the help of an E3 ubiquitin ligase. The family of isopeptidases responsible for the removal of ubiquitin from the substrate is called deubiquitination enzymes (DUBs; Millard and Wood 2006).

Attachment of a single ubiquitin moiety to a single lysine on a substrate results in monoubiquitination (Fig. 1). Monoubiquitin can be conjugated to several lysine residues on the same substrate molecule, resulting in multi-monoubiquitination. Additional ubiquitin molecules can be attached to the lysine residues in ubiquitin itself, leading to the formation of di-ubiquitin and polyubiquitin chains conjugated to a single lysine of the substrate. Although ubiquitin contains seven lysine residues, all capable of conjugating ubiquitin, Lys48- and Lys63-linked chains are the most abundant. The majority of published studies suggest that Lys48-linked chains serve as the recognition signal by the proteasome and target proteins for proteasomal degradation (Pickart and Fushman 2004). In contrast, Lys63-linked ubiquitin chains do not target proteins to proteasome but mediate interactions with protein machineries involved in endocytic trafficking, inflammatory response, protein translation, and DNA repair (Pickart and Fushman 2004). Similarly, it is widely accepted that monoubiquitination does not target proteins to the proteasome but serves as a molecular recognition signal in membrane trafficking, regulation of endocytic machinery, and possibly other cellular processes (Staub and Rotin 2006). Interestingly, the impairment of the ubiquitin-mediated protein degradation and proteosomal function

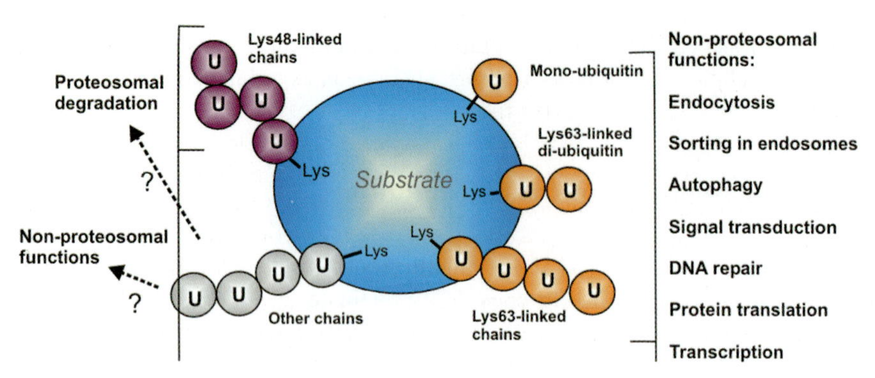

Fig. 1 Types of ubiquitin conjugation. The last residue of ubiquitin (Gly76) is covalently attached to the ε-amino group of lysines in the substrate. Substrates can be modified with a single ubiquitin molecule at single (monoubiquitination) or multiple (multi-monoubiquitination) lysine residues. Further ubiquitin conjugation to the lysine residues of the ubiquitin molecule results in the attachment of di-ubiquitin to the substrate or a substrate polyubiquitination. The main functions of monoubiquitination and the most frequently detected ubiquitin chains linked through Lys63 or Lys48 of ubiquitin are listed. Lys48- or Lys63-linked chains are shown in a "closed" or "extended" conformation, respectively, resulting in different mechanisms of recognition of these chains by ubiquitin binding domains (UBDs). Ubiquitin chains linked to other lysines of the ubiquitin have been implicated in the proteosomal and non-proteosomal processes

in neurodegenerative diseases leads to the accumulation of proteins containing mainly Lys48-linked polyubiquitin chains but also Lys63- and Lys11-linked chains (Bennett et al. 2007).

All functions of ubiquitin are accomplished through specific interactions of the ubiquitin moiety with the ubiquitin-binding domains (UBDs) found in many proteins (Hicke et al. 2005). All of the helical UBDs interact with hydrophobic Ile44 in ubiquitin, although there are several types of UBD that have different modes of recognition of mono- and poly-ubiquitin (Hurley et al. 2006). Structural studies demonstrated that Lys48-linked di-ubiquitin has a closed conformation, whereas Lys63-linked di-ubiquitin has an extended conformation, thus implying their selective recognition by different types of UBDs (Raasi et al. 2005; Varadan et al. 2004, 2005).

3 Regulation of Endocytosis of EGFR by Ubiquitination

EGFR regulates growth and survival signaling in many types of cells. EGFR signaling via the Akt pathway plays a key role in the protection of dopaminergic neurons from neurodegeneration in Parkinson's disease (Inoue et al. 2007; Iwakura et al. 2005). Binding of EGF or other ligands to the surface EGFR leads to activation of the receptor kinase and phosphorylation of C-terminal tyrosine residues, which results in recruitment of adaptor proteins and enzymes to the receptor and initiation of several signaling cascades. Activation of EGFR also causes rapid internalization of ligand-occupied EGFR through clathrin-coated pits into endosomes and subsequent efficient sorting of these complexes to the lysosome degradation pathway. Endocytosis of EGFR has a key role in the control of the intensity and duration of signaling by the receptors by down-regulating the activated EGFRs. Endocytosis is also orchestrating signaling processes by localizing EGFR and down-stream signaling effectors to various intracellular compartments. However, the molecular mechanisms of endocytosis and post-endocytic sorting of EGFR and other RTKs remain elusive.

The first clue to the mechanism of EGFR internalization came from RNA interference (RNAi) experiments in which siRNA knock-down of the GrbB2 adaptor protein demonstrated that this protein is essential for the clathrin-mediated endocytosis of EGFR. Dominant-negative mutants of Grb2 and mutation of Grb2 binding sites in EGFR reduced the internalization of EGFR. Grb2 was present in clathrin-coated pits in EGF-stimulated cells. All this evidence strongly indicated that Grb2 is important for the internalization of EGFR.

Grb2 binds to EGFR via its SH2 domain and functions as a link to bring to the receptor other proteins that are associated with the SH3 domains of Grb2 (Fig. 2). One family of proteins called *Cbls* that interact with Grb2 has been previously implicated in EGFR endocytosis and degradation, and we therefore tested the importance of Grb2-Cbl interaction in EGFR internalization. The human Cbl family of proteins consists of three isoforms, c-Cbl, Cbl-b and Cbl-c (Thien and Langdon 2001). Cbls are the E3 ubiquitin ligases. All three Cbls have an N-terminal tyrosine kinase binding (TKB) domain connected (with a linker segment) to a RING finger domain. c-Cbl and Cbl-b each have an extended C-terminal tail containing proline-rich motifs capable of binding to SH3 domains. The TKB domain directly binds to the specific phosphotyrosine-containing motifs in EGFR and other RTKs. The RING domain of the E3 ubiquitin ligase recruits an E2 enzyme and positions it so that the ubiquitin moiety can be transferred from E2 to the substrate. In our experiments, mutants of Cbl lacking Grb2 binding sites or RING domain activity have imposed a dominant-negative effect on EGFR internalization, suggesting the role of Cbl and its functional domains in EGFR internalization. This hypothesis was supported in experiments where knockdown of two Cbls (c-Cbl and Cbl-b) that interact with Grb2 by siRNA blocked internalization of EGFR.

Our studies using FRET demonstrated that the Grb2-Cbl complex is recruited to activated EGFR. The TKB domain of Cbl also directly binds to the receptor phosphorylated Tyr1045. Both direct and Grb2-mediated interactions of Cbl with the EGFR are necessary for the full ubiquitination of EGFR (Huang and Sorkin 2005; Jiang and Sorkin 2003; Levkowitz et al. 1999). This putative mechanism of dual Cbl interaction with an RTK was also demonstrated for another RTK, HGF/c-Met receptors (Peschard et al. 2001). Mutation of Tyr1045 did not affect EGFR internalization, suggesting that the direct interaction of Cbl with EGFR and full ubiquitination of the receptor are not necessary for internalization. Because the Y1045A mutant of EGFR still has residual (10–20%) ubiquitination, the question was whether this minor ubiquitination mediates internalization of EGFR.

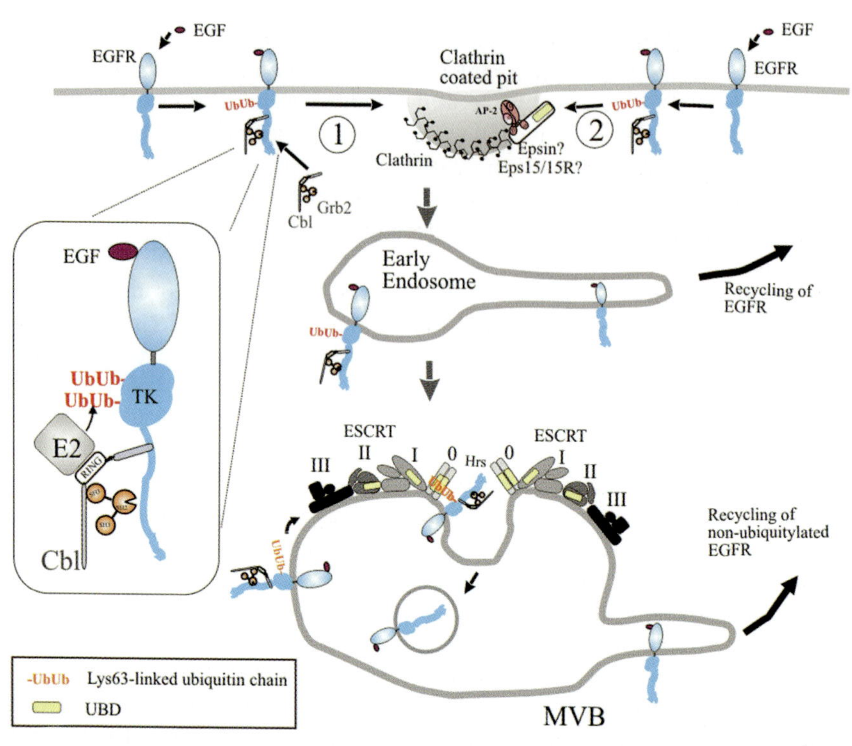

Fig. 2 Interactions of the EGF receptor leading to receptor ubiquitination and the hypothetic model of EGFR endocytosis. EGF binding activates the receptor tyrosine kinase and results in the phosphorylation of Tyr1045, Tyr1068, and Tyr1086 in the C terminus of EGFR. The SH3 domains of Grb2 are associated with the C-terminus of c-Cbl or Cbl-b. A Grb2-Cbl complex binds to the receptor by means of the interaction of the SH2 domain of Grb2 with phosphorylated Tyr1068 or Tyr1086, and the interaction of the tyrosine kinase binding (TKB) domain of c-Cbl/Cbl-b with phosphoTyr1045. Recruitment of E2 enzymes to the RING domain of Cbl results in the covalent attachment of mono-ubiquitin and poly-ubiquitin chains to the kinase domain of the receptor. EGFR is internalized via clathrin-coated pits with participation of Grb2 and Cbl by an unknown mechanism (1) or by means of the interaction of ubiquitin attached to the receptor kinase domain with the proteins containing UBD domains and located in coated pits (Eps15/Eps15R/epsin). The latter proteins can interact with the AP-2 complex or directly with clathrin. After fusion of clathrin-coated vesicles with early endosomes, EGFR can either recycle directly back to the plasma membrane or remain in the maturing endosome that acquires ESCRT complexes. Ubiquitinated receptors bind to the UBD of the ESCRT-0 complex (HRS) and eventually become trapped in the intralumenal vesicles of MVB. Non-ubiquitinated receptors can recycle back to the cell surface through the tubular extensions of MVB

To directly address the role of EGFR ubiquitination, we used mass-spectrometry analysis to map ubiquitination sites in the EGFR. Surprisingly, this analysis revealed that all the major sites of EGFR ubiquitination were located within the conserved kinase domain of the receptor (Huang et al. 2006). Additionally, in the absence of the major conjugation sites, other lysines became ubiquitinated, suggesting that EGFR ubiquitination sites were highly redundant. Importantly, quantitative mass-spectrometry analysis showed that EGFRs contained approximately 50% of mono-ubiquitin and 50% of poly-ubiquitin and that the most abundant type of polyubiquitination was the Lys63-linked chains (Huang et al. 2006).

Mutation of the major ubiquitination sites in the EGFR (lysine-to-arginine; KR mutations) had no effect on its internalization (Huang et al. 2006). However, the possibility remained that a residual cryptic ubiquitination of EGFR KR mutants was sufficient for their internalization. Therefore, in recent studies a number of other lysine residues in the EGFR kinase domain were mutated. Some lysines could not be mutated due to the loss of receptor kinase activity. However, a mutant in which 15 lysines were mutated possessed normal kinase activity but very little if any ubiquitination (about 1% of wild-type EGFR). This mutant was normally internalized, indicating that EGFR ubiquitination was not essential for internalization.

One of the multi-KR mutants, 16KR, displayed a low internalization rate. However, it was found that this mutant had reduced tyrosine kinase activity. Because tyrosine kinase activity is critical for EGFR internalization, reduced activity could explain the low rate of internalization of this mutant. However, when two major ubiquitination sites were added back by mutating two arginines back to lysines (16KR/2RK mutant), the resulting mutant was partially ubiquitinated and internalized at a rate comparable to wild-type EGFR, despite its partially reduced kinase activity. These data suggested that ubiquitination of the receptor might mediate its internalization even in the absence of the full kinase activity. Altogether, the EGFR mutagenesis experiments suggested that there were at least two redundant mechanisms of EGFR internalization through clathrin pathway. One mechanism required a full kinase activity of the receptor but did not require ubiquitination. Another mechanism utilized ubiquitination of the receptor.

4 Role of Ubiquitination in the Endosomal Sorting of EGFR

After internalization into early endosomes, receptors are either recycled back to the plasma membrane or sorted to late endosomes and lysosomes (Fig. 2). After 15–20 min of continuous EGF-induced endocytosis, EGF and EGFR accumulate in the intralumenal vesicles of multi-vesicular endosomes or bodies (MVBs) that are mostly located in the perinuclear area of the cell (McKanna et al. 1979; Miller et al. 1986). EGFRs that are incorporated into intralumenal vesicles cannot recycle. MVBs have tubular membrane extensions that are thought to be responsible for recycling of receptors not incorporated into internal vesicles (Hopkins 1992).

When the degradation rates of ubiquitination-deficient EGFR mutants were analyzed, it was found that receptor degradation was significantly decreased in all mutants of EGFR in which ubiquitination was reduced (Huang et al. 2006). Moreover, fluorescence microscopy analysis demonstrated that these mutants were inefficiently delivered to late endosomes. Finally, preliminary electron microscopy studies showed that ubiquitin-deficient EGFR mutants accumulated at the limiting membrane of MVB and in recycling endosomes whereas their incorporation into intralumenal vesicles of MVBs was significantly reduced as compared to wild-type EGFR. Therefore, ubiquitination is critical for the efficient sorting of EGFR in MVB and lysosomal targeting of the receptor.

These studies support the model whereby the ubiquitinated EGFR in endosomes interacts with the UBD of the hepatocyte growth factor receptor phosphorylation substrate (Hrs) that is associated with another UBD-containing protein, STAM1/2 (ESCRT, endosomal sorting complex required for transport, –0 complex; Bache et al. 2003; Hurley and Emr 2006). It is hypothesized that multiprotein ESCRT-I, II and III complexes surrounding cargo associated with ESCRT-0 then generate inward invagination of the limiting membrane of MVBs, thus capturing EGFR in the forming intralumenal vesicle (Babst et al. 2000; Bache et al. 2006; Bowers et al. 2006; Hurley and Emr 2006; Slagsvold et al. 2006).

Degradation of EGF and the EGFR is completely blocked by lysosomal inhibitors, suggesting that it occurs in lysosomes (Carpenter and Cohen 1976; Stoscheck and Carpenter 1984). Although the use of proteasomal inhibitors can also reduce EGFR degradation (Longva et al. 2002), these inhibitors may affect the activity of lysosomal enzymes and turnover of ESCRT proteins, or reduce the ubiquitin pool in the cell. Therefore, the effects of proteasomal inhibitors on EGFR degradation are likely indirect. The current model suggests that proteolytic enzymes are delivered to MVBs through fusion with "primary" lysosomal vesicles, which leads to the formation of mature lysosomes and proteolysis of the intralumenal content of these organelles (Miller et al. 1986).

A number of proteins have been proposed to modulate the process of EGFR targeting to the lysosome degradation pathway, mainly through affecting Cbl and Cbl-mediated ubiquitination of EGFR. Interestingly, EGFR degradation is regulated by the protein called Spartin, which is mutated in Troyer syndrome, an autosomal recessive hereditary spastic paraplegia. Thus, impaired endocytosis of EGFR or similar RTKs may underlie the pathogenesis of Troyer syndrome.

Importantly, regulation of the endocytic trafficking and stability (turnover rates) by ubiquitination is a common feature of several families of RTKs, including RTKs that are critical for the neuronal development and the survival signaling in adult neurons. For example, ubiquitination of the receptor for the nerve growth factor, TrkA, has been recently reported and implicated in the regulation of TrkA endocytosis (Arevalo et al. 2006; Geetha et al. 2005). There is disagreement as to what E3 ubiquitin ligase is involved. One study proposed that TrkA is ubiquitinated by the TRAF6 ubiquitin ligase and that this process requires the interaction of TrkA with the $p75^{NTR}$ co-receptor (Geetha et al. 2005). It is noteworthy that, similar to the EGFR, the TrkA was proposed to be polyubiqutinated by Lys63-linked chains, which was shown to be critical for endocytosis (Geetha et al. 2005). In contrast, another study claimed that TrkA is ubiquitinated by another E3 ligase, termed neuronal precursor cell expressed developmentally downregulated (NEDD4-2), which contains a HECT (homologous to E6-AP C-terminal) domain (Arevalo et al. 2006). Although the data regarding the TrkA-specific ubiquitin ligase are conflicting, both studies suggest that ubiquitination mediates endocytosis of TrkA and therefore affects signal transduction by this RTK. Examples of other RTKs that regulate survival signaling in the central nervous system and that are regulated by ubiquitination are the platelet-derived growth factor receptor (PDGFR; Mori 1993), ErbB3 and ErbB4 (Cao et al. 2007) and the insulin-like growth factor 1 receptors (Vecchione et al. 2003).

5 Regulation of DAT by Ubiquitination

Plasma membrane neurotransmitter transporters of the SLC6 family play important roles in neuronal cytotoxicity, development of neurodegenerative disorders such as Parkinson's disease, and drug abuse (Gainetdinov and Caron 2003; Gether et al. 2006). Hence, we will focus on our recent studies of one of the members of this family, DAT.

DAT is expressed in dopaminergic neurons, most of which project from the substantia nigra and ventral-tagmental area to the striatum, nucleus accumbens and prefrontal cortex. DAT functions to terminate dopamine (DA) neurotransmission via the reuptake of released DA into dopaminergic neurons. Several psychostimulants and neurotoxins, such as amphetamines, 6-hydroxydopamine (6-OHDA) and 1-methyl-4-phenyl-1,2,3,6-tetrahydropyridine (MPTP), are transported into the dopamine neuron by DAT, which can lead to dopaminergic neurodegeneration, presumably due to the accumulation of cytosolic dopamine and its oxidation into toxic dopamine-quinones (German et al. 1996; Hanrott et al. 2006; Lotharius and Brundin 2002; Sonsalla et al. 1996; Xu et al. 2005). DAT is shown to directly interact with α-synuclein, a protein involved in the development of Parkinson's disease (Lotharius et al. 2002; Lotharius and Brundin 2002), which results in reduced DAT surface expression (Lee et al. 2001).

DAT has 12 transmembrane domains and intracellular N- and C-termini (Gether et al. 2006). There are no conventional endocytosis sequence motifs in the DAT molecule. RNAi analysis showed that DAT is internalized via a clathrin-mediated pathway (Sorkina et al. 2005). Using HeLa cells expressing human DAT tagged with two epitopes at the N-terminus, we have been able to purify a sufficient amount of DAT protein to perform a mass-spectrometry analysis of purified DAT. This analysis revealed that DAT was constitutively ubiquitinated and that activation of protein kinase C (PKC) substantially increased DAT ubiquitination (Miranda et al. 2005). Furthermore, mass spectrometry also revealed the presence of Lys63-linked polyubiquitin chains in DAT. Interestingly, Western blot analysis of wild-type DAT and various lysine mutants of DAT predicted that each DAT molecule was conjugated at any given time with a single short chain of three ubiquitins.

To examine which proteins regulate PKC-induced endocytosis of DAT, we performed a large-scale RNAi screen using a reverse-transfection library of siRNAs that targeted 53 proteins implicated in endocytosis. This screen revealed that PKC-dependent DAT endocytosis required NEDD4-2 (Sorkina et al. 2006), which is an E3 ubiquitin ligase that has been implicated in the ubiquitination of various transport proteins (Miranda and Sorkin 2007). NEDD4-2 has been most well studied as an E3 ligase controlling the ubiquitination and endocytosis of ENaC channels (Staub et al. 1996). Furthermore, siRNA to NEDD4-2 dramatically decreased PKC-induced ubiquitination of DAT, suggesting that NEDD4-2 could be an E3 ligase for DAT. The NEDD4 family of proteins has a catalytic C-terminal HECT domain, the N-terminal C2 domain that binds phospholipids in a Ca^{2+}-dependent manner, and two to four WW domains that bind to the PxY (PY) motif (x is any amino acid) in the target protein (Staub and Rotin 2006). Such PY motifs are found in the C-terminal tails of various transmembrane proteins. However, a number of transporters that are regulated by NEDD4-2, including DAT, lack the PY motif. It is possible that NEDD4-2 binds indirectly to DAT, in a manner similar to that described for the IGF-1 receptor (Boehmer et al. 2006). Another possibility is that NEDD4-2 may regulate another E3 ligase that directly ubiquitinates DAT.

PKC-induced DAT ubiquitination takes place initially at the plasma membrane and continues after endocytosis. The major ubiquitination sites in the amino- and carboxyl-termini of DAT were mapped by mass spectrometry (Miranda et al. 2005). Mutagenesis of lysines in the DAT revealed that a cluster of three N-terminal lysines (Lys19, 27 and 35) is essential for PKC-dependent endocytosis of DAT (Miranda et al. 2007). PKC-induced internalization of DAT was dramatically inhibited by mutation of the ubiquitination sites (Miranda et al. 2007).

Finally, an siRNA screen revealed that the PKC-dependent internalization of DAT required the adaptor proteins epsin, Eps15, and Eps15R, which are located in clathrin-coated pits and possess UBDs (Fig. 3; Sorkina et al. 2006). Similarly, epsin and Eps15 have been recently shown to be involved in the NEDD4-2 dependent internalization of ENaC (Wang et al. 2006).

The existing methods of measuring the rate parameters of endocytic trafficking of DAT do not allow the quantification of internalization rates without the contribution of recycling. Therefore, the steps of endocytic trafficking of transporters that are regulated by ubiquitination cannot be precisely defined. Whereas several sets of data suggest that activation of PKC results in the accelerated internalization of DAT in a ubiquitin-dependent manner, it also leads to the accelerated degradation of DAT in lysosomes (Daniels and Amara 1999; Miranda et al. 2005). Therefore, it is likely that DAT ubiquitination also mediates the sorting of DAT to the degradation versus recycling pathway. As described above for the EGFR model (Fig. 2), this sorting probably involves incorporation of the transporters in the intralumenal vesicles of MVB. The observations of the co-localization of DAT with HRS in endosomes (Miranda et al. 2005; Sorkina et al. 2003) and the detection of DAT inside MVBs in DA neurons support this hypothesis (Hersch et al. 1997). It is likely that lysosomal sorting of DAT occurs mainly in the somatodendritic compartment of the dopaminergic neurons where MVBs and lysosomes are easily detected, whereas endocytic trafficking of DAT at the axonal processes in the striatum could be limited by cycling between plasma membrane and early endosomes (Fig. 3). Overall, more detailed structure-function and electron microscopy studies should be performed to characterize the role of NEDD4-2 and ubiquitination in the intracellular sorting of transporters. However, a striking similarity in the regulation of these processes among various receptor and transporter proteins is already quite evident (Miranda and Sorkin 2007).

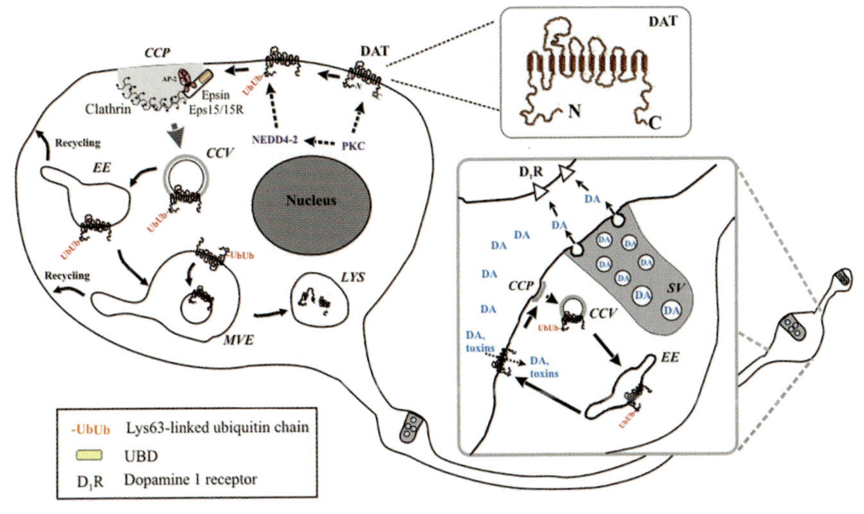

Fig. 3 Hypothetic model of endocytosis and endosomal sorting of DAT. In the somatodendritic part of DA neurons, the activation of PKC results in the NEDD4-2-mediated ubiquitination of DAT. PKC activation can facilitate the NEDD4-2-mediated ubiquitination of DAT either by phosphorylating DAT or DAT-interacting proteins or by activating NEDD4-2. Ubiquitinated DAT is recruited into clathrin-coated pits (*CCP*) by means of interaction with the UBD-containing proteins, such as Eps15/Eps15R and epsin, bound to AP-2 and clathrin in coated pits. After internalization via coated vesicles (*CCV*), DAT is sorted in early endosomes (*EE*) and MVB to lysosomes (*Lys*), presumably by a mechanism similar to that of the EGFR (Fig. 2). In the synapses of the distal axonal processes, DAT is internalized and recycled in a manner similar to that in the neuronal soma, although there is likely no sorting to late endosomes in axonal varicosities because distal axons of dopaminergic neurons lack these late endosomal compartments

6 Conclusions and Outstanding Issues

Ubiquitination has recently emerged as a critical post-translation modification that controls subcellular localization and turnover of transmembrane proteins, many of which are implicated in human neurodegenerative disease and may represent important therapeutic targets. The general consensus is that ubiquitination of the integral membrane proteins mediates the post-endocytic sorting of these proteins to lysosomes. In contrast, the role of ubiquitination in the internalization step of trafficking has been directly demonstrated only for a few endocytic cargoes in mammalian cells. The view that the regulatory functions of ubiquitination in endocytic trafficking are mediated exclusively by mono-ubiquitination has now been questioned. It is now clear that Lys63-linked polyubiquitination is the common modification of many types of transmembrane proteins. It can be proposed that, whereas monoubiquitin binds to most UBDs with low affinity, the linear conformation of Lys63 ubiquitin chains allows multivalent interactions of the same UBD-containing proteins with Lys63-polyubiqutinated cargo, thus increasing the avidity of the interaction, as compared to the interaction with mono-ubiquitin. Further investigation is needed to examine the precise role of Lys63-linked chains in endocytic trafficking.

The role of Lys63-linked polyubiquitination in neurodegenerative disease is emerging. Parkin, a protein frequently mutated in Parkinson's patients, is an E3 ubiquitin ligase that mediates formation of Lys63-ubiquitin chains, and it has been suggested that the aberrant regulation Lys63-linked polyubiquitination may result in Parkinson's disease (Doss-Pepe et al. 2005). In light of the possible role of Lys63-chains in the sorting process in the MVB, it would be interesting to investigate the relationship of the Lys63-ubiquitination and autophagy in neurons. On one hand, several studies demonstrated the important role of MVB and ESCRT complexes in autophagy (Filimonenko et al. 2007; Lee et al. 2007). These data indicate that efficient autophagic degradation requires functional MVBs and provide a possible explanation to the observed neurodegenerative phenotype seen in patients with mutations in the CHMP2B protein a part of the ESCRT III complex. On the other hand, Lys63-linked ubiquitination was found to selectively facilitate the clearance of inclusions via autophagy (Tan et al. 2008). These data indicate that Lys63-linked ubiquitin chains may represent a common modulator of inclusions biogenesis, as well as a general molecule signal targeting cargo to the autophagic system. Since autophagy has a key role in the prevention of the formation of the inclusion bodies in neurodegenerative disease, it is likely that interactions with the ESCRT complexes mediated by the Lys63-polyubiquitin chains in MVBs may be an important step that can be affected during the development of the disease.

Acknowledgements The work of the author is supported by the National Institute of Drug Abuse, National Cancer Institute and American Cancer Society.

References

Arevalo JC, Waite J, Rajagopal R, Beyna M, Chen ZY, Lee FS, Chao MV (2006) Cell survival through Trk neurotrophin receptors is differentially regulated by ubiquitination. Neuron 50:549–559.

Babst M, Odorizzi G, Estepa EJ, Emr SD (2000) Mammalian tumor susceptibility gene 101 (TSG101) and the yeast homologue, Vps23p, both function in late endosomal trafficking. Traffic 1:248–258.

Bache KG, Raiborg C, Mehlum A, Stenmark H (2003) STAM and Hrs are subunits of a multivalent ubiquitin-binding complex on early endosomes. J Biol Chem 278:12513–12521.

Bache KG, Stuffers S, Malerod L, Slagsvold T, Raiborg C, Lechardeur D, Walchli S, Lukacs GL, Brech A, Stenmark H (2006) The ESCRT-III subunit hVps24 is required for degradation but not silencing of the epidermal growth factor receptor. Mol Biol Cell 17:2513–2523.

Bennett EJ, Shaler TA, Woodman B, Ryu KY, Zaitseva TS, Becker CH, Bates GP, Schulman H, Kopito RR (2007) Global changes to the ubiquitin system in Huntington's disease. Nature 448:704–708.

Boehmer C, Palmada M, Rajamanickam J, Schniepp R, Amara S, Lang F (2006) Post-translational regulation of EAAT2 function by co-expressed ubiquitin ligase Nedd4-2 is impacted by SGK kinases. J Neurochem 97:911–921.

Bowers K, Piper SC, Edeling MA, Gray SR, Owen DJ, Lehner PJ, Luzio JP (2006) Degradation of endocytosed epidermal growth factor and virally ubiquitinated major histocompatibility complex class I is independent of mammalian ESCRTII. J Biol Chem 281:5094–5105.

Cao Z, Wu X, Yen L, Sweeney C, Carraway KL, 3rd (2007) Neuregulin-induced ErbB3 downregulation is mediated by a protein stability cascade involving the E3 ubiquitin ligase Nrdp1. Mol Cell Biol 27:2180–2188.

Carpenter G, Cohen S (1976) ^{125}I-Labeled human epidermal growth factor: binding internalization, and degradation in human fibroblasts. J Cell Biol 71:159–171.
Cirrito JR, Kang JE, Lee J, Stewart FR, Verges DK, Silverio LM, Bu G, Mennerick S, Holtzman DM (2008) Endocytosis is required for synaptic activity-dependent release of amyloid-beta in vivo. Neuron 58:42–51.
Daniels GM, Amara SG (1999) Regulated trafficking of the human dopamine transporter. Clathrin-mediated internalization and lysosomal degradation in response to phorbol esters. J Biol Chem 274:35794–35801.
DiFiglia M, Sapp E, Chase KO, Davies SW, Bates GP, Vonsattel JP, Aronin N (1997) Aggregation of huntingtin in neuronal intranuclear inclusions and dystrophic neurites in brain. Science 277:1990–1993.
Doss-Pepe EW, Chen L, Madura K (2005) Alpha-synuclein and parkin contribute to the assembly of ubiquitin lysine 63-linked multiubiquitin chains. J Biol Chem 280:16619–16624.
Filimonenko M, Stuffers S, Raiborg C, Yamamoto A, Malerod L, Fisher EM, Isaacs A, Brech A, Stenmark H, Simonsen A (2007) Functional multivesicular bodies are required for autophagic clearance of protein aggregates associated with neurodegenerative disease. J Cell Biol 179:485–500.
Gainetdinov RR, Caron MG (2003): Monoamine transporters: from genes to behavior. Annu Rev Pharmacol Toxicol 43:261–284.
Geetha T, Jiang J, Wooten MW (2005) Lysine 63 Polyubiquitination of the nerve growth factor receptor TrkA directs internalization and signaling. Mol Cell 20:301–312.
German DC, Nelson EL, Liang CL, Speciale SG, Sinton CM, Sonsalla PK (1996) The neurotoxin MPTP causes degeneration of specific nucleus A8, A9 and A10 dopaminergic neurons in the mouse. Neurodegeneration 5:299–312.
Gether U, Andersen PH, Larsson OM, Schousboe A (2006) Neurotransmitter transporters: molecular function of important drug targets. Trends Pharmacol Sci 27:375–383.
Hanrott K, Gudmunsen L, O'Neill MJ, Wonnacott S (2006) 6-hydroxydopamine-induced apoptosis is mediated via extracellular auto-oxidation and caspase 3-dependent activation of protein kinase Cdelta. J Biol Chem 281:5373–5382.
Hersch SM, Yi H, Heilman CJ, Edwards RH, Levey AI (1997) Subcellular localization and molecular topology of the dopamine transporter in the striatum and substantia nigra. J Comp Neurol 388:211–227.
Hershko A, Ciechanover A (1992) The ubiquitin system for protein degradation. Annu Rev Biochem 61:761–807.
Hicke L, Dunn R (2003) Regulation of membrane protein transport by ubiquitin and ubiquitin-binding proteins. Annu Rev Cell Dev Biol 19:141–172.
Hicke L, Riezman H (1996) Ubiquitination of a yeast plasma membrane receptor signals its ligand-stimulated endocytosis. Cell 84:277–287.
Hicke L, Schubert HL, Hill CP (2005) Ubiquitin-binding domains. Nature Rev Mol Cell Biol 6:610–621.
Hopkins CR (1992) Selective membrane protein trafficking: vectorial flow and filter [see comments]. Trends Biochem Sci 17:27–32.
Huang F, Sorkin A (2005) Growth factor receptor binding protein 2-mediated recruitment of the RING domain of Cbl to the epidermal growth factor receptor is essential and sufficient to support receptor endocytosis. Mol Biol Cell 16:1268–1281.
Huang F, Kirkpatrick D, Jiang X, Gygi S, Sorkin A (2006) Differential regulation of EGF receptor internalization and degradation by multiubiquitination within the kinase domain. Mol Cell 21:737–748.
Hurley JH, Emr SD (2006) The ESCRT complexes: structure and mechanism of a membrane-trafficking network. Annu Rev Biophys Biomol Struct 35:277–298.
Hurley JH, Lee S, Prag G (2006) Ubiquitin-binding domains. Biochem J 399:361–372.
Inoue H, Lin L, Lee X, Shao Z, Mendes S, Snodgrass-Belt P, Sweigard H, Engber T, Pepinsky B, Yang L, Beal MF, Mi S, Isacson O (2007) Inhibition of the leucine-rich repeat protein LINGO-1 enhances survival, structure, and function of dopaminergic neurons in Parkinson's disease models. Proc Natl Acad Sci USA 104:14430–14435.

Iwakura Y, Piao YS, Mizuno M, Takei N, Kakita A, Takahashi H, Nawa H (2005) Influences of dopaminergic lesion on epidermal growth factor-ErbB signals in Parkinson's disease and its model: neurotrophic implication in nigrostriatal neurons. J Neurochem 93:974–983.

Jiang X, Sorkin A (2003) Epidermal growth factor receptor internalization through clathrin-coated pits requires Cbl RING finger and proline-rich domains but not receptor polyubiquitylation. Traffic 4:529–543.

Kolling R, Hollenberg CP (1994) The ABC-transporter Ste6 accumulates in the plasma membrane in a ubiquitinated form in endocytosis mutants. Embo J 13:3261–3271.

Lee FJ, Liu F, Pristupa ZB, Niznik HB (2001) Direct binding and functional coupling of alpha-synuclein to the dopamine transporters accelerate dopamine-induced apoptosis. Faseb J 15:916–926.

Lee JA, Beigneux A, Ahmad ST, Young SG, Gao FB (2007) ESCRT-III dysfunction causes autophagosome accumulation and neurodegeneration. Curr Biol 17:1561–1567.

Levkowitz G, Waterman H, Ettenberg SA, Katz M, Tsygankov AY, Alroy I, Lavi S, Iwai K, Reiss Y, Ciechanover A, Lipkowitz S, Yarden Y (1999) Ubiquitin ligase activity and tyrosine phosphorylation underlie suppression of growth factor signaling by c-Cbl/Sli-1 [In Process Citation]. Mol Cell 4:1029–1040.

Longva KE, Blystad FD, Stang E, Larsen AM, Johannessen LE, Madshus IH (2002) Ubiquitination and proteasomal activity is required for transport of the EGF receptor to inner membranes of multivesicular bodies. J Cell Biol 156:843–854.

Lotharius J, Brundin P (2002) Pathogenesis of Parkinson's disease: dopamine, vesicles and alpha-synuclein. Nature Rev Neurosci 3:932–942.

Lotharius J, Barg S, Wiekop P, Lundberg C, Raymon HK, Brundin P (2002) Effect of mutant alpha-synuclein on dopamine homeostasis in a new human mesencephalic cell line. J Biol Chem 277:38884–38894.

Lowe J, Blanchard A, Morrell K, Lennox G, Reynolds L, Billett M, Landon M, Mayer RJ (1988) Ubiquitin is a common factor in intermediate filament inclusion bodies of diverse type in man, including those of Parkinson's disease, Pick's disease, and Alzheimer's disease, as well as Rosenthal fibres in cerebellar astrocytomas, cytoplasmic bodies in muscle, and mallory bodies in alcoholic liver disease. J Pathol 155:9–15.

Mayer RJ, Lowe J, Lennox G, Doherty F, Landon M (1989) Intermediate filaments and ubiquitin: a new thread in the understanding of chronic neurodegenerative diseases. Prog Clin Biol Res 317:809–818.

McKanna JA, Haigler HT, Cohen S (1979) Hormone receptor topology and dynamics: morphological analysis using ferritin-labeled epidermal growth factor. Proc Natl Acad Sci USA 76:5689–5693.

Millard SM, Wood SA (2006) Riding the DUBway: regulation of protein trafficking by deubiquitylating enzymes. J Cell Biol 173:463–468.

Miller K, Beardmore J, Kanety H, Schlessinger J, Hopkins CR (1986) Localization of epidermal growth factor (EGF) receptor within the endosome of EGF-stimulated epidermoid carcinoma (A431) cells. J. Cell Biol. 102:500–509.

Miranda M, Sorkin A (2007) Regulation of receptors and transporters by ubiquitination: new insights into surprisingly similar mechanisms. Mol Interv 7:157–167.

Miranda M, Wu CC, Sorkina T, Korstjens DR, Sorkin A (2005) Enhanced ubiquitylation and accelerated degradation of the dopamine transporter mediated by protein kinase C. J Biol Chem. 280:35617–35624.

Miranda M, Dionne KR, Sorkina T, Sorkin A (2007) Three ubiquitin conjugation sites in the amino terminus of the dopamine transporter mediate protein kinase C-dependent endocytosis of the transporter. Mol Biol Cell 18:313–323.

Mori S, Rönnstrand L, Yokote K, Engström A, Courtneidge SA, Claesson-Welsh L, Heldin CH (1993) Identification of two juxtamembrane autophosphorylation sites in the PDGF beta-receptor; involvement in the interaction with Src family tyrosine kinases. Embo J. 12:2257–2264.

Mukhopadhyay D, Riezman H (2007) Proteasome-independent functions of ubiquitin in endocytosis and signaling. Science 315:201–205.

Peschard P, Fournier TM, Lamorte L, Naujokas MA, Band H, Langdon WY, Park M (2001) Mutation of the c-Cbl TKB domain binding site on the Met receptor tyrosine kinase converts it into a transforming protein. Mol Cell 8:995–1004.

Pickart CM, Fushman D (2004): Polyubiquitin chains: polymeric protein signals. Curr Opin Chem Biol 8:610–616.

Raasi S, Varadan R, Fushman D, Pickart CM (2005) Diverse polyubiquitin interaction properties of ubiquitin-associated domains. Nature Struct Mol Biol 12:708–714.

Ross CA, Pickart CM (2004) The ubiquitin-proteasome pathway in Parkinson's disease and other neurodegenerative diseases. Trends Cell Biol 14:703–711.

Slagsvold T, Pattni K, Malerod L, Stenmark H (2006) Endosomal and non-endosomal functions of ESCRT proteins. Trends Cell Biol 16:317–326.

Sonsalla PK, Jochnowitz ND, Zeevalk GD, Oostveen JA, Hall ED (1996) Treatment of mice with methamphetamine produces cell loss in the substantia nigra. Brain Res 738:172–175.

Sorkina T, Doolen S, Galperin E, Zahniser NR, Sorkin A (2003) Oligomerization of dopamine transporters visualized in living cells by fluorescence resonance energy transfer microscopy. J Biol Chem. 278:28274–28283.

Sorkina T, Hoover BR, Zahniser NR, Sorkin A (2005) Constitutive and protein kinase C-induced internalization of the dopamine transporter is mediated by a clathrin-dependent mechanism. Traffic 6:157–170.

Sorkina T, Miranda M, Dionne KR, Hoover BR, Zahniser NR, Sorkin A (2006) RNA interference screen reveals an essential role of Nedd4-2 in dopamine transporter ubiquitination and endocytosis. J Neurosci 26:8195–8205.

Staub O, Dho S, Henry P, Correa J, Ishikawa T, McGlade J, Rotin D (1996) WW domains of Nedd4 bind to the proline-rich PY motifs in the epithelial Na+ channel deleted in Liddle's syndrome. EMBO J. 15:2371–2380.

Staub O, Rotin D (2006) Role of ubiquitylation in cellular membrane transport. Physiol Rev 86:669–707.

Stoscheck CM, Carpenter G (1984) "Down-regulation" of EGF receptors: Direct demonstration of receptor degradation in human fibroblasts. J Cell Biol 98:1048–1053.

Tan JM, Wong ES, Kirkpatrick DS, Pletnikova O, Ko HS, Tay SP, Ho MW, Troncoso J, Gygi SP, Lee MK, Dawson VL, Dawson TM, Lim KL (2008) Lysine 63-linked ubiquitination promotes the formation and autophagic clearance of protein inclusions associated with neurodegenerative diseases. Human Mol Genet 17:431–439.

Thien CB, Langdon WY (2001) Cbl: many adaptations to regulate protein tyrosine kinases. Nature Rev Mol Cell Biol 2:294–307.

Varadan R, Assfalg M, Haririnia A, Raasi S, Pickart C, Fushman D (2004) Solution conformation of Lys63-linked di-ubiquitin chain provides clues to functional diversity of polyubiquitin signaling. J Biol Chem 279:7055–7063.

Varadan R, Assfalg M, Raasi S, Pickart C, Fushman D (2005) Structural determinants for selective recognition of a Lys48-linked polyubiquitin chain by a UBA domain. Mol Cell 18:687–698.

Vecchione A, Marchese A, Henry P, Rotin D, Morrione A (2003) The Grb10/Nedd4 complex regulates ligand-induced ubiquitination and stability of the insulin-like growth factor I receptor. Mol Cell Biol 23:3363–3372.

Wang H, Traub LM, Weixel KM, Hawryluk MJ, Shah N, Edinger RS, Perry CJ, Kester L, Butterworth MB, Peters KW, Kleyman TR, Frizzell RA, Johnson JP (2006) Clathrin-mediated endocytosis of the epithelial sodium channel: ROLE OF EPSIN. J Biol Chem 281:14129–14135.

Xu W, Zhu JP, Angulo JA (2005) Induction of striatal pre- and postsynaptic damage by methamphetamine requires the dopamine receptors. Synapse 58:110–121.

Zweifel LS, Kuruvilla R, Ginty DD (2005) Functions and mechanisms of retrograde neurotrophin signalling. Nature Rev Neurosci 6:615–625.

The Sortilin-Related Receptor SORL1 is Functionally and Genetically Associated with Alzheimer's Disease

Ekaterina Rogaeva, Yan Meng, Joseph H. Lee, Richard Mayeux, Lindsay A. Farrer, and Peter St George-Hyslop(✉)

Abstract The recycling of the amyloid precursor protein (APP) from the cell surface via the endocytic pathways plays a key role in the generation of amyloid β-peptide (Aβ), the accumulation of which is thought to be central to the pathogenesis of Alzheimer's disease (AD). Inherited variants in the SORL1 neuronal sorting receptor have been reproducibly associated with late-onset AD. These variants occur in intronic sequences and may regulate tissue-specific expression of SORL1, which directs trafficking of APP into recycling pathways. When SORL1 is under-expressed, APP is sorted into Aβ-generating compartments. These data lead to the conclusion that inherited or acquired changes in SORL1 gene expression or function are mechanistically involved in causing AD.

1 Introduction

The accumulation of amyloid β-peptide (Aβ), a neurotoxic proteolytic derivative of the amyloid precursor protein (APP) is a central event in the pathogenesis of Alzheimer's disease (AD; Mattson 2004). Thus, inherited variants in the amyloid precursor protein (APP; Goate et al. 1991), presenilin 1 (PS1; Sherrington et al. 1995) presenilin 2 (PS2; Rogaev et al. 1995) and apolipoprotein E (APOE) all cause Aβ accumulation in the brain (Saunders et al. 1993; Bales et al. 1997). The generation of Aβ occurs in several subcellular compartments, but a principle location is during the re-entry and recycling of APP from the cell surface via the endocytic pathway (Golde et al. 1992; Haass and Selkoe 1993; Bayer et al. 2003;

P. St George-Hyslop
Centre for Research in Neurodegenerative Diseases, Departments of Medicine, Laboratory Medicine and Pathobiology, Medical Biophysics, University of Toronto, and Toronto Western Hospital Research Institute, Toronto, Ontario, Canada;
Cambridge Institute for Medical Research and Dept of Clinical Neurosciences, University of Cambridge, Wellcome Trust/MRC Building, Addenbrookes Hospital, Hills Road, Cambridge UK CB2 0XY, E-mail: p.hyslop@utoronto.ca

P. St. George-Hyslop et al. (eds.) *Intracellular Traffic and Neurodegenerative Disorders*, Research and Perspectives in Alzheimer's Disease,

Kinoshita et al. 2003; Vetrivel et al. 2005). We reasoned that inherited variants in these pathways might modulate APP processing and thereby affect risk for AD. This concept is supported by prior reports that 1) the expression of several candidate proteins within these pathways (e.g., SORL1; Scherzer et al. 2004; VPS35; Small et al. 2005) is reduced in AD brain tissue; and 2) reductions in the expression of some of these proteins is associated with increased Aβ production (Andersen et al. 2005; Small et al. 2005; Offe et al. 2006). However, it is unclear whether these changes are causal or are simply reactive to the AD process.

To address this question, we investigated genetic associations between AD and single nucleotide polymorphisms (SNPs) in selected members of the vacuolar protein sorting (VPS) gene family, including VPS35 (16q12); VPS26 (10q21); sortilin - SORT1 (1p21-p13); sortilin-related VPS10 containing receptors - SORCS1 (10q23-q25), SORCS2 (4p16), and SORCS3 (10q23-q25); and sortilin-related receptor, low density lipoprotein receptor class A repeats-containing - SORL1 (11q23-q24; also identified as LR11 or SORLA). The inheritance of SNPs from these genes was explored in six independent datasets that had sufficient power to detect modest gene effects ($\lambda s = 1.5$; see reference for details (Rogaeva et al. 2007).

2 SNPs in SORL1 are Associated with Late-Onset AD

This survey failed to uncover any significant allelic associations with VPS26, VPS35, SORCS3, or SORT1. However, six SNPs in two clusters at the 5′ and 3′ ends of the SORL1 gene showed significant association with AD in at least one discovery dataset and also in at least one replication dataset (Fig. 1; $0.0031 \leq p \leq 0.014$). More importantly, for five of these SNPs, the association was observed with identical alleles in the discovery and replication datasets. Thus, AD was associated with the "C," "G" and "C" alleles at SNPs 8, 9 and 10, respectively, in three datasets, whereas AD was associated with the "G" and "T" alleles at SNPs 19 and 23, respectively, in two datasets. Haplotypic analyses using the sliding window method (Lin et al. 2004) and a window size of three contiguous SNPs confirmed the single SNP analyses, demonstrating replicated haplotypic associations in two regions of SORL1 in different datasets.

3 Subsequent Replications in other Datasets

These results have now been tested in eight independent reports on at least seven non-overlapping datasets comprised of AD-affected individuals (the TGEN datasets are partially overlapping). One dataset (Li et al. 2008), comprised of cases and controls drawn from multiple ethnic origins across Canada and UK (GenADA), failed to show any association at any SNP in SORL1 (but also failed to show significant association at any other locus except APOE, suggesting that this study was likely to have

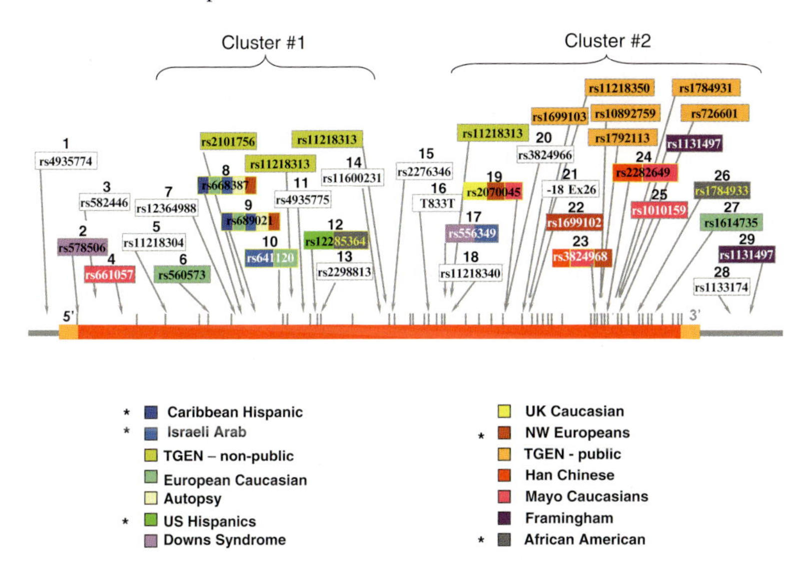

Fig. 1 Top panel: Genomic map of SORL1 gene showing the location of SNPs genotyped in the original study (Rogaeva et al. 2007). Orange bars represent the 5′UTR and 3′UTR; red bar represents intragenic regions; vertical bars represent each of the 48 exons; SNPs 1, 28 and 29 are located in extragenic intervals. The backgrounds of the flags for each SNP name are coloured to depict the dataset generating the positive result. Multiple colours reflect significant results in multiple datasets. White = no associations reported. Bottom panel: Colour codes for the ethnic origins of the datasets reported to date. * - datasets in the original publication (Rogaeva et al. 2007)

been confounded by allelic and/or non-allelic heterogeneity amongst its component ethnic groups). The remaining seven independent reports generated nominally significant associations, and the associated SNPs tended to cluster into the same two regions (Fig. 1 and Table 1; Lee et al. 2007a, 2008; Meng et al. 2007; Tan et al. 2007; Bettens et al. 2008; Li et al. 2008; Webster et al. 2008).

In addition, associations with SORL1 have been reported with risk for AD in subjects with Down's syndrome (Lee et al. 2007b) and with cognitive decline in the Framingham cohort study of normal aging ($p = 3 \times 10^{-6}$; Seshadri et al. 2007).

Finally, the same SORL1 variants previously associated with AD have also now been shown to be associated with cerebral spinal fluid (CSF) endophenotypes and with brain magnetic resonance imaging (MRI) and neuropathological endophenotypes of neurodegenerative and cerebrovascular diseases. Specifically, in a study of CSF Aβ levels, a three-marker SORL1 haplotype consisting of SNP19 T-allele, SNP21 G-allele and SNP23 A-allele (T/G/A) was associated with reduced Aβ42 CSF levels in AD patients ($p = 0.003$; Kolsch et al. 2008). In a separate study of clinically diagnosed AD cases assessed by MRI- and autopsy-confirmed cases, SNPs 16–18 were associated with AD in the clinical cohort (global $p = 0.031$; Cuenco et al. 2008). SNPs 8–10 were associated with fewer white matter hyper-intensities (WMH) in both the clinical ($p = 0.0005$) and autopsy ($p = 0.02$) series. In addition, general cerebral atrophy and hippocampal/mesial temporal atrophy were associated

Table 1 Summary of all studies on SORL1 in AD as of June 2008.

Author	Journal	# Cases/Controls	Ethnicity	SNP IDs	SNP ID in Rogaeva*	p-value
Beetens	Human Mutation 2008	550/634	Caucasian	rs560573	6	0.011
				rs668387	8	0.028
				rs689021	9	0.029
				rs641129	10	0.022
				rs1614735	27	0.001
Lee et al	Neurology 2007	103/93	Autopsy	rs668387	8	0.015
				rs689021	9	0.017
Webster	Neurodegen Dis 2007	664/422	Caucasian	rs2101756	8–9	0.019
			(TGEN non-public)	rs11218313	10–11	0.020
				rs626885	14–15	0.036
				rs7131432	17–18	0.039
Li et al	Neurobiol Dis 2007	998/1033	Caucasian	rs2070045	19	0.035
		343/346	UK Cohort	rs2282649	24	0.022
Lee et al	Arch Neurol 2007	178/194	US Hispanic	rs12285364	12	0.029
		88/158	African American	rs12285364	12	0.016
				rs1784933	26	0.018
		30/76	Caucasian	rs3824966	20	0.025
Tan et al	Neurobiol Aging 2007	223/263	Chinese	rs1699102	22	0.067
				rs3824968	23	0.042
				rs2282649	24	0.064
Meng et al	Neuroreport 2007	1044/364	Caucasian	rs1699103	20–21	0.045
			(TGEN public)	rs11218350	20–21	0.040
				rs10892759	22–23	0.021
				rs1792113	22–23	0.041
				rs726601	24–25	0.038
				rs1784931	24–25	0.049
Li et al	Arch Neurol 2008	753/736	Caucasian	none	none	none
Lee et al	Neurosci Lett 2007	208	NY Downs Synd Caucasian	rs556349	17	0.047
Seshadri et al	BMC Med Genet 2007	1705	Framingham	rs726601	24–25	8.2×10^{-4}
			(Cognition)	rs1131497	29	3.2×10^{-6}

with markers from the 5′ region, including SNPs 21–26, in both datasets (Cuenco et al. 2008). Examination of specific 3-SNP haplotypes from these two regions in the autopsy-confirmed AD cases showed association of white matter disease with SNPs 8–10 and association of hippocampal atrophy and parenchymal vascular lesions with SNPs 22–24. Of note, the same SNP 8–10 haplotype (CGC) was associated with decreased WMH in AD cases in the clinical ($p = 0.0005$) and autopsy ($p = 0.02$) samples. The observation that SORL1 variants are associated both with AD itself and with disease-relevant endophenotypes, such as CSF Aβ levels, brain imaging changes and cognitive impairment in aging, supports both the hypothesis that multiple regions of the SORL1 gene are functionally important and the hypothesis that selected individual SORL1 variants may have different effects on SORL1 expression. These different effects may selectively alter SORL1 activity in individual cell types (e.g., neurons versus endothelial cells) and/or in particular regions of the brain.

4 Cell Biology of SORL1

The SNPs and haplotypes tested to date are unlikely to be the actual AD-causing variants. Sequencing of the exons and intron-exon boundaries in carriers of the disease-associated haplotypes at SNPs 8–10 or SNPs 22–24, and investigation of SORL1 splice forms recovered by RT-PCR, both failed to identify any potentially pathogenic sequence variants that were significantly enriched in AD cases over controls. It is therefore likely that the associations reflect the presence of pathogenic variants within the intronic sequences nearby the intervals containing SNPs 8–10 and 22–24. We have speculated that these putative intronic variants might modulate cell type-specific transcription or translation of SORL1 in neurons of carriers of the AD-associated haplotypes, but directly testing this hypothesis is difficult (Rogaeva et al. 2007). The view that primary reductions in SORL1 might be causally linked to AD is supported by previous reports that SORL1 expression is reduced in neurons in some cases of sporadic AD (where the cause was unknown) but not in PS1-mutant familial AD (FAD; where the disease arises from mutations in PS1; Scherzer et al. 2004; Dodson et al. 2006).

The conclusion that SORL1 might be involved in the pathogenesis of AD is also supported by cell biological experiments that demonstrate that SORL1 specifically regulates APP trafficking into the endocytic or recycling pathways. Thus, SORL1 physically interacts with the APP holoprotein (Fig. 2) and with the VPS35 (which drives cargo selection in the retromer via VPS10-containing proteins like SORL1; Seaman 2005). These protein-protein interactions are specific because SORL1 does not bind to other Type 1 membrane proteins (e.g., BACE1; He et al. 2005; or VPS26, which links VPS35 to the other structural elements of the retromer; Seaman 2005). SORL1 also does not bind to APP fragments produced by α, β- or γ-secretase cleavage. Furthermore, over-expression of SORL1, which would be predicted to divert APP holoprotein into the retromer recycling pathway, results in decreased Aβ

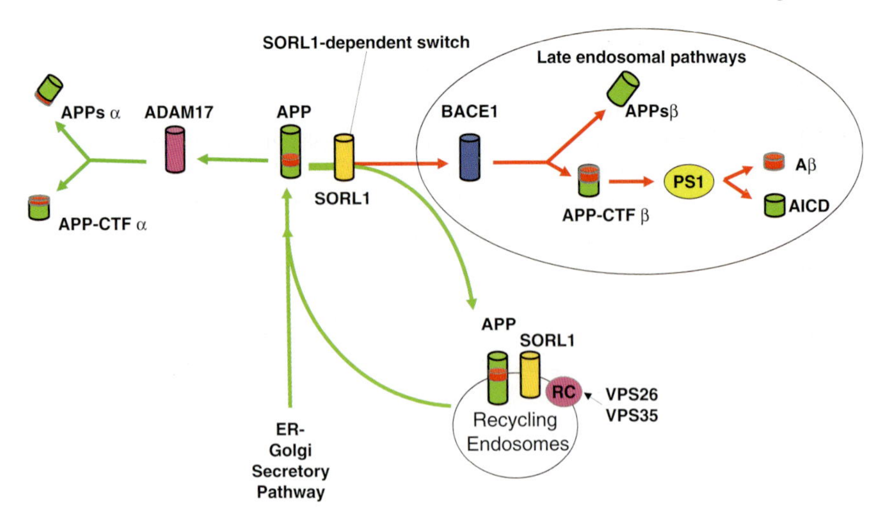

Fig. 2 Diagram of APP processing pathways. APP holoprotein is synthesized in the endoplasmic reticulum (ER) and Golgi. Proteolytic cleavage through the Aβ peptide domain by ADAM17 and other α-secretase enzymes generates N-terminal soluble APPsα and membrane-bound APP-CTFα fragments. Sequential cleavage by BACE1 (β-secretase) generates N-terminal APPsβ and membrane-bound APP-CTFβ fragments. The latter undergo presenilin-dependent γ-secretase cleavage to generate Aβ and amyloid intracellular domain (AICD). SORL1 binds both APP holoprotein and VPS35 (not shown) and acts as a sorting receptor for APP holoprotein. Absence of SORL1 switches APP holoprotein away from the retromer recycling pathway and instead directs APP into the β-secretase cleavage pathway, increasing APPsβ production, and then into the γ-secretase cleavage pathway to generate Aβ. Blockade of the retromer complex (RC) by inhibiting retromer complex proteins such as VPS26 or VPS35, or Golgi-localized gamma-ear-containing ARF-binding (GGA) adaptor proteins, has a similar effect, also increasing APPsβ and Aβ production

production (82 % of control value, $p < 0.05$). Conversely, siRNA suppression of SORL1 expression, which results in deflection of APP holoprotein away from the recycling retromer pathway and into the late endosome-lysosome pathway, causes (1) over-production of the APPsβ ectodomain and (2) over-production of Aβ by the subsequent γ-secretase cleavage of the APP C-terminal stub generated by BACE1 (Aβ40 = 189% of control; Aβ42 = 202% of control, $p < 0.001$; Andersen et al. 2005; Offe et al. 2006; Rogaeva et al. 2007).

5 Discussion

Taken together, the genetic and cell biological results from an increasing number of independent groups suggest that genetic and/or environmentally specified changes in SORL1 expression or function are causally linked to the pathogenesis of AD and have a modest effect on risk for this disease. The precise identity of the genetic effectors in SORL1 remains to be determined. However, the results described here imply that (1) there are several different allelic variants in distinct genomic regions of the SORL1 gene in different populations, (2) these variants are likely to be in intronic regulatory sequences that might govern cell-type or tissue-specific expression of SORL1, and (iii) these variants affect disease risk by altering the physiological role of SORL1 in the processing of APP holoprotein.

The observations that (1) no single SORL1 SNP or haplotype is associated with risk for AD in all datasets and (2) some datasets fail to show any association contrast sharply with APOE [where there is an association of AD with a single APOE allele (i.e., APOE $\varepsilon 4$) in most datasets (Farrer et al. 1997)]. However, it is important to note that the association of disease with a single allele in all datasets (i.e., an APOE $\varepsilon 4$-like association) is not a universal observation for either complex or monogenic diseases (Pritchard and Cox 2002). Thus, the occurrence of pathogenic mutations across multiple domains of disease genes (i.e., allelic heterogeneity), and the absence of these variants in some datasets (i.e., locus heterogeneity) are not unusual in either monogenic or complex traits (Owen et al. 2005; Vermeire and Rutgeerts 2005). In addition, several points affirmatively mitigate concerns that the association between SORL1 and AD is spurious. First, the association was initially identified using conservative family-based association tests, which are less sensitive to confounding due to population stratification (Rabinowitz and Laird 2000). Second, the results have now been replicated in multiple unrelated datasets drawn from ethnically different origins

The work summarized here argues that (1) variants in SORL1 increase risk for AD and for selected endophenotypes, (2) the reduction in SORL1 expression previously observed in AD is likely to be a primary and pathogenic event, and (3) SORL1 plays a key physiological role in the subcellular sorting of APP holoprotein.

Acknowledgements Supported by grants from The Wellcome Trust, Howard Hughes Medical Institute, Canadian Institutes of Health Research, Ontario Research Fund, Alzheimer Association of Ontario (PHStGH and ER); National Institutes of Health R37AG15473 and PO1AG07232 (RM) R01-AG09029, R01-AG25259, R01-AG17173, P30-AG13846 (LAF).

References

Andersen OM, Reiche J, Schmidt V, Gotthardt M, Spoelgen R, Behlke J, von Arnim CA, Breiderhoff T, Jansen P, Wu X, Bales KR, Cappai R, Masters CL, Gliemann J, Mufson EJ, Hyman BT, Paul SM, Nykjaer A, Willnow TE (2005) Neuronal sorting protein-related receptor sorLA/LR11 regulates processing of the amyloid precursor protein. Proc Natl Acad Sci USA 102: 13461–13466.

Bales KR, Verina T, Dodel RC, Du Y, Altsteil L, Bender M, Hyslop P, Johnstone EM, Little SP, Cummins DJ, Piccardo P, Ghetti B, Paul SM (1997) Lack of apolipoprotein E dramatically reduces amyloid beta-peptide deposition. Nature Genet 17: 254–256.

Bayer TA, Schafer S, Simons A, Kemmling A, Kamer T, Tepest R, Eckert A, Schussel K, Eikenberg O, Sturchler-Pierrat C, Abramowski D, Staufenbiel M, Multhaup G (2003) Dietary Cu stabilizes brain superoxide dismutase 1 activity and reduces amyloid Abeta production in APP23 transgenic mice. Proc Natl Acad Sci USA 100: 14187–14192.

Bettens K, Brouwers N, Engelborghs S, De Deyn PP, Van Broeckhoven C, Sleegers K (2008) SORL1 is genetically associated with increased risk for late-onset Alzheimer disease in the Belgian population. Human Mutat 29: 769–770.

Cuenco KT, Lunetta KL, Baldwin CT, McKee AC, Guo J, Cupples LA, Green RC, St George-Hyslop PH, Chui H, DeCarli C, Farrer LA, Group MS (2008) Distinct variants in SORL1 are associated with cerebrovascular and neurodegenerative changes related to Alzheimer disease. Arch Neurol, in press.

Dodson SE, Gearing M, Lippa CF, Montine TJ, Levey AI, Lah JJ (2006) LR11/SorLA expression is reduced in sporadic Alzheimer disease but not in familial Alzheimer disease. J Neuropathol Exp Neurol 65: 866–872.

Farrer LA, Cupples LA, Haines JL, Hyman B, Kukull WA, Mayeux R, Myers RH, Pericak-Vance MA, Risch N, van Duijn CM (1997) Effects of age, sex, and ethnicity on the asscoiation between apolipoprotein E genotype and Alzheimer's Disease. A meta-analysis. APOE and Alzheimer's Disease Meta Analysis consortium. JAMA 278: 1349–1356.

Goate AM, Chartier-Harlin M-C, Mullan M, Brown J, Crawford F, Fidani L, Guiffra L, Haynes A and Hardy JA (1991). Segregation of a missense mutation in the amyloid precursor protein gene with familial Alzheimer disease. Nature 349: 704–706.

Golde TE, Estus S, Younkin LH, Selkoe DJ, Younkin SG (1992) Processing of the amyloid protein precursor to potentially amyloidogenic derivatives. Science 255: 728–730.

Haass C, Selkoe DJ (1993) Cellular processing of beta-amyloid precursor protein and the genesis of amyloid beta-peptide. Cell 75: 1039–1042.

He X, Li F, Chang WP, Tang J (2005) GGA proteins mediate the recycling pathway of memapsin 2 (BACE). J Biol Chem 280: 11696–11703.

Kinoshita A, Fukumoto H, Shah T, Whelan CM, Irizarry MC, Hyman BT (2003) Demonstration by FRET of BACE interaction with the amyloid precursor protein at the cell surface and in early endosomes. J Cell Sci 116: 3339–3346.

Kolsch H, Jessen F, Wiltfang J, Lewczuk P, Dichgans M, Kornhuber J, Frolich L, Heuser I, Peters O, Schulz JB, Schwab SG, Maier W (2008) Influence of SORL1 gene variants: Association with CSF amyloid-beta products in probable Alzheimer's disease. Neurosci Lett 440: 68–71

Lee JH, Cheng R, Schupf N, Manly J, Lantigua R, Stern Y, Rogaeva E, Wakutani Y, Farrer L, St George-Hyslop P, Mayeux R (2007a) The association between genetic variants in SORL1 and Alzheimer disease in an urban, multiethnic, community-based cohort. Arch Neurol 64: 501–506.

Lee JH, Chulikavit M, Pang D, Zigman WB, Silverman W, Schupf N (2007b) Association between genetic variants in sortilin-related receptor 1 (SORL1) and Alzheimer's disease in adults with Down syndrome. Neurosci Lett 425: 105–109.

Lee JH, Cheng R, Honig LS, Vonsattel JP, Clark L, Mayeux R (2008) Association between genetic variants in SORL1 and autopsy-confirmed Alzheimer disease. Neurology 70: 887–889.

Li H, Wetten S, Li L, St Jean PL, Upmanyu R, Surh L, Hosford D, Barnes MR, Briley JD, Borrie M, Coletta N, Delisle R, Dhalla D, Ehm MG, Feldman HH, Fornazzari L, Gauthier S, Goodgame N, Guzman D, Hammond S, Hollingworth P, Hsiung GY, Johnson J, Kelly DD, Keren R, Kertesz A, King KS, Lovestone S, Loy-English I, Matthews PM, Owen MJ, Plumpton M, Pryse-Phillips W, Prinjha RK, Richardson JC, Saunders A, Slater AJ, St George-Hyslop PH, Stinnett SW, Swartz JE, Taylor RL, Wherrett J, Williams J, Yarnall DP, Gibson RA, Irizarry MC, Middleton LT, Roses AD (2008) Candidate single-nucleotide polymorphisms from a genomewide association study of Alzheimer disease. Arch Neurol 65: 45–53.

Li Y, Rowland C, Catanese J, Morris J, Lovestone S, O'Donovan MC, Goate A, Owen M, Williams J, Grupe A (2008) SORL1 variants and risk of late-onset Alzheimer's disease. Neurobiol Dis 29: 293–296.

Lin S, Chakravarti A, Cutler DJ (2004) Exhaustive allelic transmission disequilibrium tests as a new approach to genome-wide association studies. Nature Genet 36: 1181–1188.

Mattson MP (2004) Pathways towards and away from Alzheimer's disease. Nature 430: 631–639.

Meng Y, Lee JH, Cheng R, St George-Hyslop P, Mayeux R, Farrer LA (2007) Association between SORL1 and Alzheimer's disease in a genome-wide study. Neuroreport 18: 1761–1764.

Offe K, Dodson SE, Shoemaker JT, Fritz JJ, Gearing M, Levey AI, Lah JJ (2006) The lipoprotein receptor LR11 regulates amyloid beta production and amyloid precursor protein traffic in endosomal compartments. J Neurosci 26: 1596–1603.

Owen MJ, Craddock N, O'Donovan MC (2005). Schizophrenia: genes at last? Trends Genet 21: 518–525.

Pritchard JK, Cox NJ (2002) The allelic architecture of human disease genes: common disease-common variantor not? Human Mol Genet 11: 2417–2423.

Rabinowitz D, Laird N (2000) A unified approach to adjusting association tests for population admixture with arbitrary pedigree structure and arbitrary missing marker information. Human Hered 50: 211–223.

Rogaev EI, Sherrington R, Rogaeva EA, Levesque G, Ikeda M, Liang Y, Chi H, Lin C, Holman K, Tsuda T, Mar L, Sorbi S, Nacmias B, Piacentini S, Amaducci L, Chumakov I, Cohen D, Lannfelt L, Fraser PE, Rommens JM, St George-Hyslop P (1995) Familial Alzheimer's disease in kindreds with missense mutations in a novel gene on chromosome 1 related to the Alzheimer's Disease type 3 gene. Nature 376: 775–778.

Rogaeva E, Meng Y, Lee JH, Gu Y, Kawarai T, Zou F, Katayama T, Baldwin CT, Cheng R, Hasegawa H, Chen F, Shibata N, Lunetta KL, Pardossi-Piquard R, Bohm C, Wakutani Y, Cupples LA, Cuenco KT, Green RC, Pinessi L, Rainero I, Sorbi S, Bruni A, Duara R, Friedland RP, Inzelberg R, Hampe W, Bujo H, Song YQ, Andersen OM, Willnow TE, Graff-Radford N, Petersen RC, Dickson D, Der SD, Fraser PE, Schmitt-Ulms G, Younkin S, Mayeux R, Farrer LA, St George-Hyslop P (2007) The neuronal sortilin-related receptor SORL1 is genetically associated with Alzheimer disease. Nature Genet 39: 168–177.

Saunders A, Strittmatter WJ, Schmechel S, St George-Hyslop P, Pericak-Vance M, Joo SH, Rosi BL, Gusella JF, Crapper-McLachlan D, Growden J, Alberts MJ, Hulette C, Crain B, Goldgaber D, Roses AD (1993) Association of Apoliprotein E allele e4 with the late-onset familial and sporadic Alzheimer Disease. Neurology 43: 1467–1472.

Scherzer CR, Offe K, Gearing M, Rees HD, Fang G, Heilman CJ, Schaller C, Bujo H, Levey AI, Lah JJ (2004) Loss of apolipoprotein E receptor LR11 in Alzheimer disease. Arch Neurol 61: 1200–1205.

Seaman MN (2005) Recycle your receptors with retromer. Trends Cell Biol 15: 68–75.

Seshadri S, DeStefano AL, Au R, Massaro JM, Beiser AS, Kelly-Hayes M, Kase CS, D'Agostino RB, Sr., Decarli C, Atwood LD, Wolf PA (2007) Genetic correlates of brain aging on MRI and cognitive test measures: a genome-wide association and linkage analysis in the Framingham Study. BMC Med Genet 8 Suppl 1: S15.

Sherrington R, Rogaev E, Liang Y, Rogaeva E, Levesque G, Ikeda M, Chi H, Lin C, Holman K, Tsuda T, Mar L, Fraser P, Rommens JM, St George-Hyslop P (1995) Cloning of a gene bearing missense mutations in early onset familial Alzheimer's disease. Nature 375: 754–760.

Small SA, Kent K, Pierce A, Leung C, Kang MS, Okada H, Honig L, Vonsattel JP, Kim TW (2005) Model-guided microarray implicates the retromer complex in Alzheimer's disease. Ann Neurol 58: 909–919.

Tan EK, Lee J, Chen CP, Teo YY, Zhao Y, Lee WL (2007) SORL1 haplotypes modulate risk of Alzheimer's disease in Chinese. Neurobiol Aging 28: 1361–1366.

Vermeire S, Rutgeerts P (2005) Current status of genetics research in inflammatory bowel disease. Genes Immun 6: 637–645.

Vetrivel KS, Cheng H, Kim SH, Chen Y, Barnes NY, Parent AT, Sisodia SS, Thinakaran G (2005) Spatial segregation of gamma -secretase and substrates in distinct membrane domains. J Biol Chem 27: 25892–25900.

Webster JA, Myers AJ, Pearson JV, Craig DW, Hu-Lince D, Coon KD, Zismann VL, Beach T, Leung D, Bryden L, Halperin RF, Marlowe L, Kaleem M, Huentelman MJ, Joshipura K, Walker D, Heward CB, Ravid R, Rogers J, Papassotiropoulos A, Hardy J, Reiman EM, Stephan DA (2008) Sorl1 as an Alzheimer's disease predisposition gene? Neurodegener Dis 5: 60–64.

Regulation of Transport and Processing of Amyloid Precursor Protein by the Sorting Receptor SORLA

Thomas E. Willnow(✉), Michael Rohe, Anne-Sophie Carlo, and Vanessa Schmidt

Abstract Trafficking of the amyloid precursor protein (APP) through intracellular compartments where the various secretase activities reside is a regulatory step that determines APP processing fates and is likely to play a role in pathological processes leading to Alzheimer's disease (AD). SORLA is a member of a novel class of intracellular sorting proteins and acts as a neuronal receptor for APP, controlling transport and proteolytic processing of the precursor protein into amyloidogenic and non-amyloidogenic products. Substantial experimental evidence from epidemiological, cell biological and animal studies points to a model whereby SORLA determines trafficking of APP into cellular compartments less favorable for processing. Consequently, high levels of SORLA expression are associated with reduced APP processing rates, whereas low levels of the protein, as in patients with sporadic AD, may predispose to accelerated APP breakdown and enhanced senile plaque formation. This article discusses the molecular mechanisms that govern SORLA-mediated APP transport and processing and their potential relevance for neurodegenerative processes.

1 Introduction

In 1996, Petersen and colleagues and Yamakazi et al. independently identified a novel gene product from human and rabbit tissues that they termed sorting protein-related receptor with low-density lipoprotein (LDL) receptor class A repeats (SORLA; Jacobsen et al. 1996) or LDL receptor relative with 11 binding repeats (LR11; Yamazaki et al. 1996), respectively. At that time, the race had been on to find novel members of a newly discovered gene family of LDL receptor-related receptors

T.E. Willnow
Max-Delbrueck-Center for Molecular Medicine, Robert-Roessle-Str. 10, D-13125 Berlin Germany, E-mail: willnow@mdc-berlin.de

P. St. George-Hyslop et al. (eds.) *Intracellular Traffic and Neurodegenerative Disorders*, Research and Perspectives in Alzheimer's Disease,

(LRPs), a group of multifunctional endocytic receptors with structural homology to the LDL receptor (Nykjaer and Willnow 2002).

Indeed, SORLA did harbor some of the unifying motifs that characterize LRPs, including complement (or A)-type repeats, epidermal growth factor-type repeats, and β-propellers (Fig. 1). However, the SORLA polypeptide also contained additional structural features not seen in LRPs before. These motifs included six fibronectin type III elements as well as a 700-amino acid module that was initially identified in the vacuolar protein sorting 10 protein (VPS10p), a sorting receptor in *Yeast* that directs carboxypeptidase Y from the trans-Golgi network (TGN) to the vacuole (the *Yeast* homologue of the lysosome; Marcusson et al. 1994).

Subsequent cloning efforts uncovered even more eukaryotic gene products characterized by the VPS10p domain, including the pro-nerve growth factor receptor sortilin (Nykjaer et al. 2004; Petersen et al. 1997), the *Hydra* head activator binding protein (Hampe et al. 1999, 2000), and the neuronal receptors, SORCS1, SORCS2, and SORCS3 (Hampe et al. 2001). Because all of the latter proteins lack any obvious homology to the LDL receptor, it is now generally accepted that SORLA represents the founding member of a separate class of VPS10p-domain receptors that are both structurally and functionally distinct from LRPs. In particular, a predominant localization of VPS10p-domain receptors in intracellular compartments suggests a function in protein sorting rather than endocytosis. This assumption was supported by early evidence that some of the receptors interacted with GGA-1 and -2 (Golgi-localizing, γ-adaptin ear homology domain, ARF-interacting proteins; Jacobsen et al. 2002; Nielsen et al. 2001), which are adaptors that direct proteins between trans-Golgi network (TGN) and early endosomes (Bonifacino 2004; Bonifacino and Traub 2003; Zhu et al. 2001). Although the regular trafficking route of SORLA had not been established at that time, related members of the gene family were shown to move to the cell surface, internalize once, and thereafter shuttle between endosomal compartments and TGN (Nielsen et al. 2001).

In the mammalian organism, SORLA is found primarily in neurons of the central and peripheral nervous system (Hermans-Borgmeyer et al. 1998; Motoi et al. 1999), where it localizes to the plasma membrane and to various intracellular compartments, such as Golgi and early endosomes (Hampe et al. 2000; Jacobsen et al. 2001). In the brain, the receptor is widely expressed throughout the cortex, hippocampus, and cerebellum (Hermans-Borgmeyer et al. 1998; Motoi et al. 1999). Furthermore, significant expression has been documented in non-neuronal tissues, including kidney, testis, and liver (Jacobsen et al. 1996; Yamazaki et al. 1996). The functions that SORLA may have at these sites remain largely unclear, but some studies suggested a role as cellular receptor for apolipoprotein E (Taira et al. 2001), platelet-derived growth factor BB (Gliemann et al. 2004), or glia-derived neurotrophic factor (Westergaard et al. 2004). Also, a possible role as regulator of smooth muscle cell migration has been considered (Zhu et al. 2004).

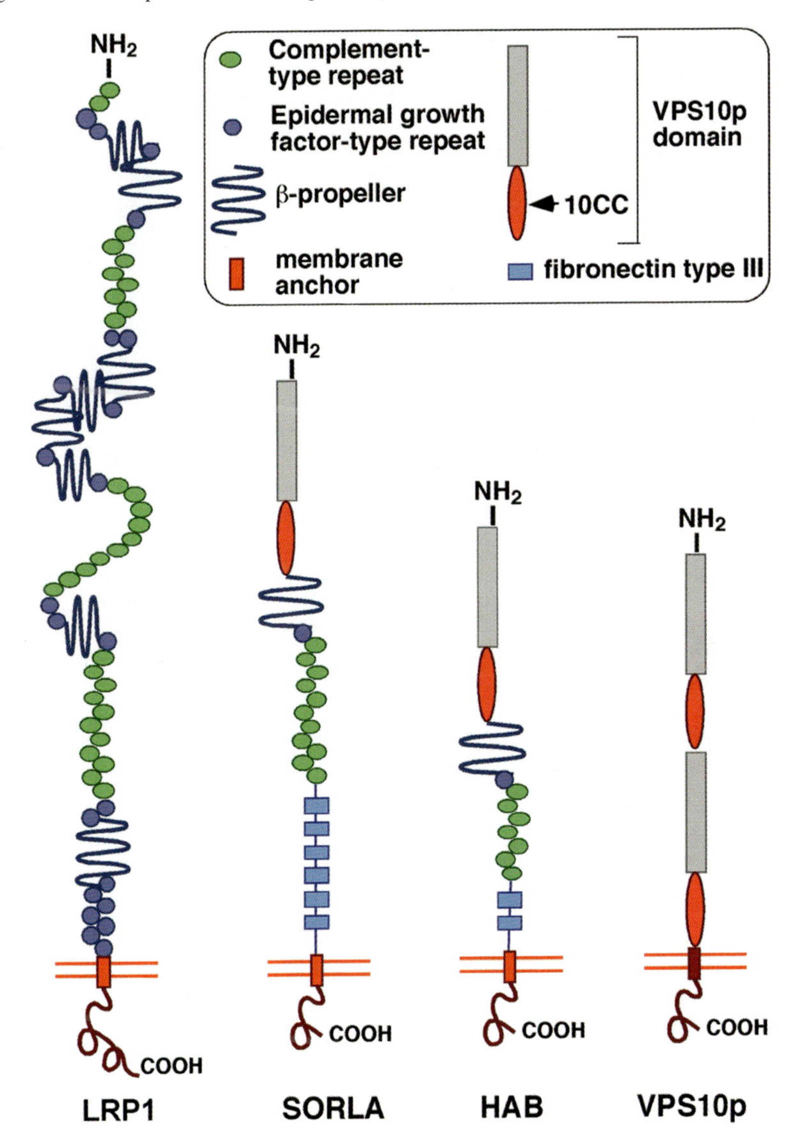

Fig. 1 Structural organization of SORLA and related receptors. The figure depicts common structural motifs in SORLA, low-density lipoprotein receptor related protein 1 (LRP1), head activator binding protein (HAB), and vacuolar protein sorting 10 protein (VPS10p). Complement-type repeats represent ligand binding sites, whereas β-propellers are involved in release of receptor-bound macromolecules in endocytic compartments. The significance of other protein domains in the context of receptor function (e.g., 10CC, fibronectin type III) is unclear at present

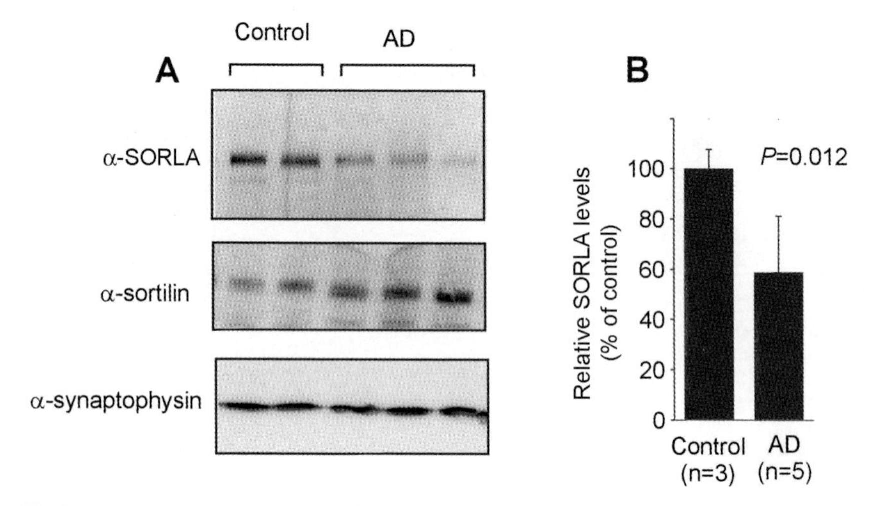

Fig. 2 Loss of SORLA expression in patients with the sporadic form of Alzheimer's disease (AD). (**A**) Brain specimens from three individuals with sporadic AD and two control subjects were subjected to Western blot analysis using antibodies directed against SORLA, sortilin, and the neuronal marker, synaptophysin. Loss of expression in AD is specific for SORLA and not seen for sortilin or synaptophysin. (**B**) Densitometric scanning of replicate Western blots (as in A) indicates a 40% reduction in SORLA levels in AD patients compared to healthy controls

2 Expression of SORLA is Lost in Patients with Sporadic Alzheimer's Disease

A major breakthrough in functional characterization of SORLA came with an observation made by Scherzer et al. (2004), who used gene expression profiling to uncover a reduction of SORLA mRNA levels in lymphoblasts from Alzheimer's disease (AD) patients. Almost complete absence of receptor expression in individuals with AD was confirmed by Western blot and immunohistological analyses of brain autopsies (Andersen et al. 2005; Dodson et al. 2006; Scherzer et al. 2004; Fig. 2). Intriguingly, a reduction of SORLA levels was specifically documented in patients suffering from late-onset AD but not in individuals with familial forms of the disease (Dodson et al. 2006). These observations linked SORLA through a yet unknown activity to neurodegenerative processes. In particular, the data suggested low levels of SORLA as a primary cause of sporadic AD rather than a secondary consequence of the neuronal cell loss in AD patients.

3 SORLA Acts as a Neuronal Sorting Receptor for Amyloid Precursor Protein

What might be the molecular mechanism whereby SORLA affects AD processes in the brain? Based on its structural homology to sorting receptors, a similar function for SORLA in neuronal transport of amyloid precursor protein (APP) was proposed

(Andersen et al. 2005). APP follows a complex, intracellular trafficking pathway that influences processing to either a soluble fragment sAPPα (non-amyloidogenic) or to sAPPβ and the insoluble amyloid β-peptide (Aβ), the principal component of senile plaques (De Strooper and Annaert 2000). The rate of Aβ production is considered a major risk factor for onset of AD (De Strooper and Annaert 2000). En route through the secretory pathway to the cell surface, most newly synthesized APP molecules are cleaved into sAPPα by α-secretase whereas some precursor molecules are re-internalized from the plasma membrane and delivered to endocytic compartments for β-secretase (and subsequent γ-secretase) processing into sAPPβ and Aβ (De Strooper and Annaert 2000; see model in Fig. 6). Accordingly, the intracellular transport and localization of APP are crucial determinants of APP processing and Aβ production. Yet considerable controversy exists regarding the mechanisms that govern intracellular transport of the precursor protein.

A decisive role for SORLA in the intracellular trafficking of APP has now been confirmed in a number of studies that demonstrated direct interaction between the sorting receptor and APP in neurons. Binding to SORLA was shown for all three major APP isoforms: APP_{770}, APP_{751}, and the neuronal variant APP_{695} (Fig. 3). Interaction involves binding sites in the extracellular as well as in the cytoplasmic tail region of both proteins (Andersen et al. 2005, 2006; Spoelgen et al. 2006). In particular, fine-mapping identified a binding epitope within the cluster of complement-type repeats in SORLA that forms a 1:1 stoichiometric complex with the carbohydrate-linked domain of APP (Andersen et al. 2006). Interaction of the two proteins mainly occurs in *late*-Golgi/TGN and in early endocytic compartments, as shown by confocal immunocytochemistry and fluorescence lifetime imaging microscopy (Andersen et al. 2005, 2006; Spoelgen et al. 2006). Functional interaction results in impaired transition of APP through the Golgi, effectively reducing the number of precursor molecules that reach the plasma membrane. In contrast, SORLA does not affect the rate of internalization of APP from the cellular surface, in line with a presumed function in intracellular (but not endocytic) transport processes (Spoelgen et al. 2006).

4 SORLA Impairs APP Processing

The central role of the Golgi in APP metabolism is well appreciated as it represents the major site of APP concentration in the cell (Caporaso et al. 1994). More importantly, initial processing of APP by α- and β-secretases is intimately associated with a *post*-Golgi compartment and requires efficient transit of the precursor through this organelle (Haass et al. 1993; Yamazaki et al. 1995). Thus, disrupting Golgi transition of APP blocks processing (Khvotchev and Sudhof 2004; Peraus et al. 1997), whereas phorbol ester treatment that enhances membrane shunt from the TGN to the plasma membrane increases APP processing (Xu et al. 1995). Because SORLA delayed APP exit from the Golgi, these observations suggested a mode of action whereby SORLA-mediated sequestration of APP in the Golgi might impair access

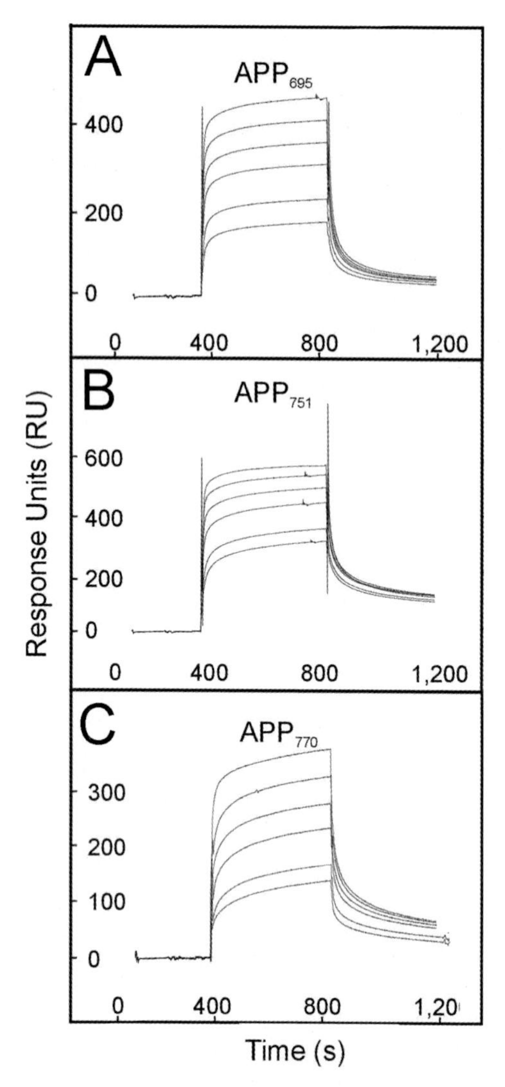

Fig. 3 SORLA interacts with all major APP variants. Surface plasmon resonance analysis demonstrates interaction of APP_{695} (**A**), APP_{751} (**B**), and APP_{770} (**C**) with the recombinant extracellular domain of SORLA immobilized on the sensor chip surface. A concentration series of APP variants at 0.1, 0.2, 0.5, 1.0, 2.0, 5.0 μM was applied

of the precursor to the *post*-Golgi compartments where proteolytic processing by secretases proceeded. Intriguingly, such a protective role for SORLA in the prevention of APP processing was indeed confirmed in a number of studies in cultured cell lines, including CHO, HEK293, as well as neuronal N2A and SH-SY5Y cells, that demonstrated a significant reduction in APP processing when SORLA was overexpressed (Fig. 4A, B; Andersen et al. 2005; Offe et al. 2006; Spoelgen et al.

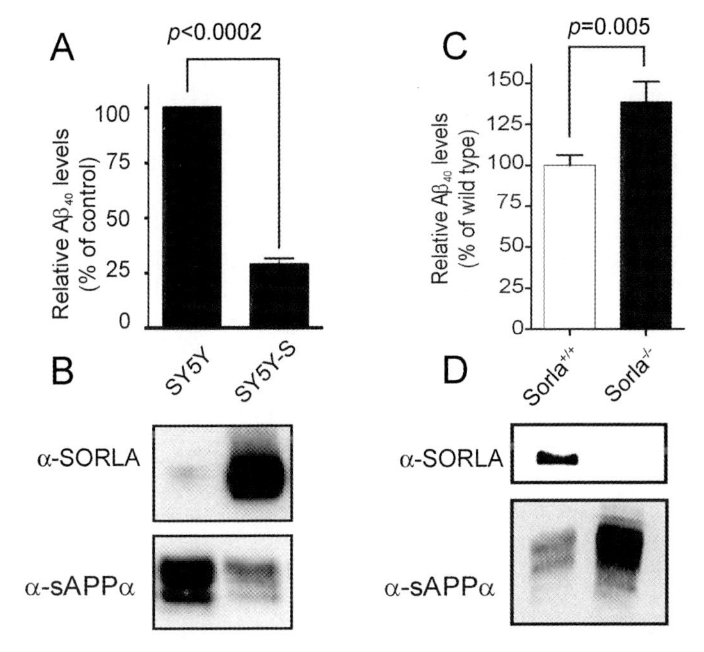

Fig. 4 Levels of SORLA expression affect APP processing rates. Western blot analysis (**B**) and ELISA (**A**) were used to quantify levels of SORLA, soluble APPα (sAPPα), and Aβ in parental neuronal cell line SH-SY5Y (SY5Y) or SH-SY5Y cells stably overexpressing SORLA (SY5Y-S). sAPPα and Aβ levels were significantly reduced in SY5Y-S compared to parental SY5Ycells. Detection of SORLA and sAPPα (**D**) and of Aβ (**C**) in hippocampal extracts from wild type (*Sorla*$^{+/+}$) and SORLA-deficient mice (*Sorla*$^{-/-}$) indicates increased levels of APP processing products in receptor-deficient animals

2006). The reduction in processing efficiency affected both amyloidogenic and non-amyloidogenic pathways. Detailed analysis of APP processing products indicated that the receptor exerted its inhibitory effect via blockade of α- and β-secretase activities (Schmidt et al. 2007).

Recently, the significance of SORLA for APP processing was also confirmed by studies in mice with targeted *Sorla* gene disruption. In this mouse model, loss of receptor expression coincided with significantly higher levels of Aβ and sAPPα in the brain compared to control animals (Fig. 4C, D), similar to the situation seen in AD patients who lack receptor expression (Andersen et al. 2005).

5 SORLA Activity Requires Interaction with GGA and PACS-1

Similar to the mode of action of other sorting proteins (such as sortilin or mannose 6-phosphate receptors), functional expression of SORLA involves interaction with cytosolic adaptor proteins. A number of cellular mechanisms target proteins to and from the Golgi/TGN, including interaction with sorting adaptors GGA and PACS-1 (Bonifacino and Traub 2003; Ghosh and Kornfeld 2004). Binding of GGA-1 and -2 to a tetrapeptide motif DVPM in the tail of SORLA had been demonstrated before, but the functional relevance for receptor trafficking and activity had not been investigated (Jacobsen et al. 2002). In addition, an acidic cluster that may serve as binding site for PACS-1 is also present in the cytoplasmic receptor domain. To dissect regulatory elements in SORLA that convey Golgi/TGN targeting, Schmidt et al. (2007) generated mutant forms of the receptor that lacked the presumed GGA (SORLA^{gga}) or PACS-1 (SORLA^{acidic}) binding motifs, or the entire cytoplasmic domain (SORLA$^{\Delta cd}$). When trafficking of these mutants was compared to the wild type receptor in neuronal and non-neuronal cell types, both SORLA^{acidic} and SORLA$^{\Delta cd}$ failed to localize to the Golgi but were accumulated at the cell surface. In contrast, SORLA^{gga} was partially able to reside in the Golgi but unable to efficiently recycle from endocytic compartments back to the TGN. Aberrant trafficking of SORLA variants profoundly changed the processing pattern of APP co-expressed with the mutants. Thus, trapping of APP in recycling compartments (as with SORLA^{gga}) stimulated processing by α-secretase (Fig. 5A) whereas shunt to the cell surface (as with SORLA^{acidic} and SORLA$^{\Delta cd}$) massively accelerated cleavage by β-/γ-secretases, likely by enhancing delivery of APP molecules into the endocytic pathway (Fig. 5B). Intriguingly, ß-site APP-cleaving enzyme 1 (BACE-1) has also been identified as a target of GGA-mediated trafficking in cells (von Arnim et al. 2004). In line with observations that SORLA and BACE-1 localize in close proximity in Golgi compartments of cultured neurons (Spoelgen et al. 2006), the above finding suggests the existence of a supramolecular protein complex composed of adaptors, sorting receptors, and secretases, as well as the substrate APP (through interaction with SORLA), that may be central to the transport and processing of the precursor protein.

6 Conclusion

Currently, all available experimental evidence points to a central role for SORLA in control of APP transport to and from the Golgi/TGN (Fig. 6; Andersen and Willnow 2006). Newly synthesized APP molecules may first encounter the receptor when they enter the Golgi on their way through the secretory pathway to the cell surface (step 1 in Fig. 6). SORLA-mediated retention of APP in this organelle requires the activity of PACS-1 and delays entry of APP molecules into the non-amyloidogenic (step 2) and amyloidogenic (step 3) processing pathways. Consistent with this model, high levels of SORLA expression further reduce APP processing rates (Fig. 4A, B), whereas low levels of receptor activity, as in mouse models of SORLA deficiency, accelerate Golgi transit and increase processing efficiency (Fig. 4C, D).

As well as APP, some SORLA molecules may reach the cell surface from where they internalize via clathrin-coated pit endocytosis. From the early endocytic

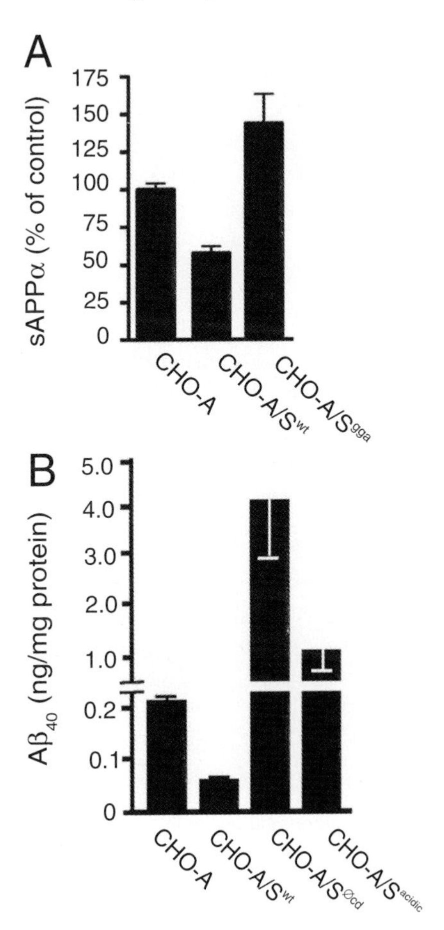

Fig. 5 Abnormal trafficking of SORLA alters APP processing rates. (**A**) Determination by semi-quantitative Western blots of sAPPα levels in parental Chinese hamster ovary cells expressing APP only (CHO-A) or APP with the wild type (CHO-A/S^{wt}) or the GGA mutant form (CHO-A/S^{gga}) of SORLA. (**B**) Quantification by ELISA of $A\beta_{40}$ levels in the medium of CHO cells expressing human APP only (CHO-A), APP with the wild type (CHO-A/S^{wt}), the PACS mutant (CHO-A/S^{acidic}) or the tail-less form (CHO-A/$S^{\Delta cd}$) of SORLA

compartments, SORLA molecules recycle back to *trans*-Golgi/TGN through the action of GGAs (step 4). Endocytosis and recycling of SORLA do not affect trafficking of APP in the endocytic compartments (Schmidt et al. 2007).

An additional regulatory mechanism in SORLA trafficking that is not fully understood may involve the retromer, a multimeric protein complex responsible for retrograde trafficking of proteins from late endosomes/lysosomes to the Golgi (reviewed in Seaman 2004, 2005). VPS35 is the main component of the retromer and is known to bind to *Yeast* VPS10p (Nothwehr et al. 1999). This observation led

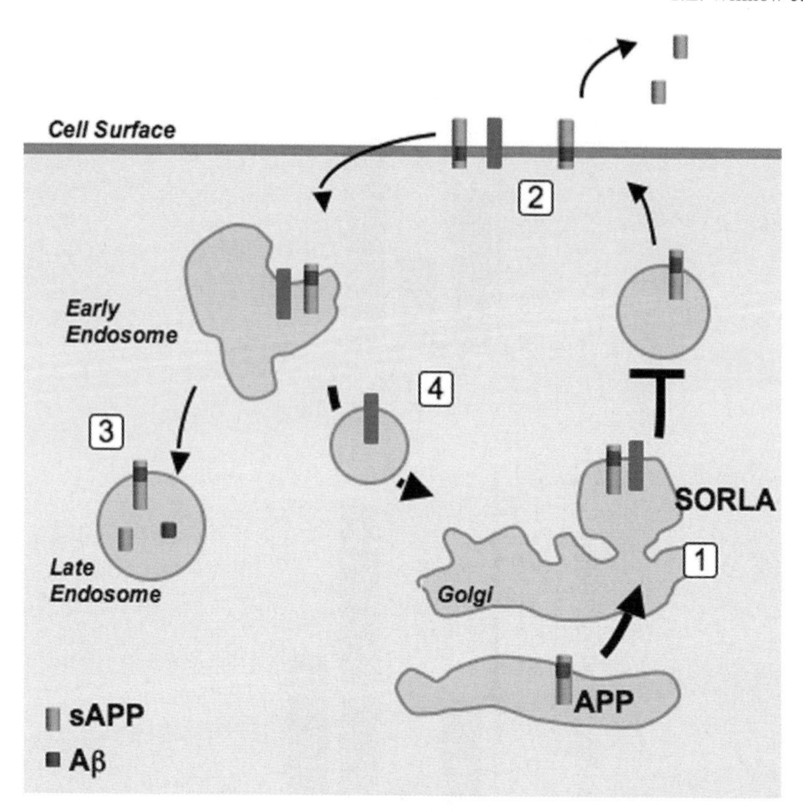

Fig. 6 SORLA function in APP transport and processing. Typically, nascent APP molecules traverse the Golgi (1) en route to the plasma membrane where some are cleaved by α-secretase to sAPPα (non-amyloidogenic pathway) (2). Non-processed precursors internalize from the cell surface and traffic from early to late endosomes for cleavage into sAPPβ and Aβ (amyloidogenic pathway) (3). SORLA acts as a sorting receptor that traps APP in the Golgi, thereby reducing the number of precursor molecules that can be processed in *post*-Golgi compartments (1). Retention of SORLA (and of APP) in the Golgi entails functional interaction of SORLA with PACS-1. Recycling of internalized SORLA molecules from the early endocytic compartment back to the Golgi/TGN requires the activity of GGA (4)

to the suggestion that a similar interaction between retromer and SORLA may also take place in mammalian cells (Small and Gandy 2006). Reducing retromer activity by selective depletion of individual protein components from cells (e.g., VPS35) leads to an increase in Aβ secretion, whereas overexpression of VPS35 reduces Aβ levels (Small et al. 2005), similar to the effects of SORLA on APP processing (Andersen et al. 2005).

Future studies should provide more insights into the molecular details of the SORLA trafficking machinery in neurons that seems central to the cellular catabolism of APP. They may even uncover new molecular targets to modulate this pathway in patients with AD and to interfere with pathological processes in this devastating disorder.

Acknowledgements Work in the authors' laboratory described here was funded by grants from the German Research Foundation, the Alzheimer Forschung Initiative e.V., the European Commission FP6, and the American Health Assistance Foundation.

References

Andersen OM, Willnow TE (2006). Lipoprotein receptors in Alzheimer's disease. Trends Neurosci 29: 687–694.

Andersen OM, Reiche J, Schmidt V, Gotthardt M, Spoelgen R, Behlke J, von Arnim CA, Breiderhoff T, Jansen P, Wu X, Bales KR, Cappai R, Masters CK. Gliemann J, Mufson EJ, Hyman BT, Paul SM, Nykjaer A, Willnow TE (2005) Neuronal sorting protein-related receptor sorLA/LR11 regulates processing of the amyloid precursor protein. Proc Natl Acad Sci USA 102: 13461–13466.

Andersen OM, Schmidt V, Spoelgen R, Gliemann J, Behlke J, Galatis D, McKinstry WJ, Parker MW, Masters CL, Hyman BT, Cappai R, Willnow TE (2006) Molecular dissection of the interaction between amyloid precursor protein and its neuronal trafficking receptor SorLA/LR11. Biochemistry 45: 2618–2628. Bonifacino JS (2004) The GGA proteins: adaptors on the move. Nature Rev Mol Cell Biol 5: 23–32.

Bonifacino JS, Traub LM (2003) Signals for sorting of transmembrane proteins to endosomes and lysosomes. Annu Rev Biochem 72: 395–447.

Caporaso GL, Takei K, Gandy SE, Matteoli M, Mundigl O, Greengard P, De Camilli P (1994) Morphologic and biochemical analysis of the intracellular trafficking of the Alzheimer beta/A4 amyloid precursor protein. J Neurosci 14: 3122–3138.

De Strooper B, Annaert W (2000) Proteolytic processing and cell biological functions of the amyloid precursor protein. J Cell Sci 113 (Pt 11): 1857–1870.

Dodson SE, Gearing M, Lippa CF, Montine TJ, Levey AI, Lah JJ (2006) LR11/SorLA expression is reduced in sporadic Alzheimer disease but not in familial Alzheimer disease. J Neuropathol Exp Neurol 65: 866–872.

Ghosh P, Kornfeld S (2004) The GGA proteins: key players in protein sorting at the trans-Golgi network. Eur J Cell Biol 83: 257–262.

Gliemann J, Hermey G, Nykjaer A, Petersen CM, Jacobsen C, Andreasen PA (2004) The mosaic receptor sorLA/LR11 binds components of the plasminogen activating system and PDGF-BB similarly to low density lipoprotein receptor-related protein (LRP1) but mediates slow internalization of bound ligand. Biochem J 381: 203–212.

Haass C, Hung AY, Schlossmacher MG, Teplow DB, Selkoe DJ (1993) beta-Amyloid peptide and a 3-kDa fragment are derived by distinct cellular mechanisms. J Biol Chem 268: 3021–3024.

Hampe W, Urny J, Franke I, Hoffmeister-Ullerich SA, Herrmann D, Petersen CM, Lohmann J, Schaller HC (1999) A head-activator binding protein is present in hydra in a soluble and a membrane-anchored form. Development 126: 4077–4086.

Hampe W, Riedel IB, Lintzel J, Bader CO, Franke I, Schaller HC (2000) Ectodomain shedding, translocation and synthesis of SorLA are stimulated by its ligand head activator. J Cell Sci 113: 4475–4485.

Hampe W, Rezgaoui M, Hermans-Borgmeyer I, Schaller HC (2001) The genes for the human VPS10 domain-containing receptors are large and contain many small exons. Human Genet 108: 529–536.

Hermans-Borgmeyer I, Hampe W, Schinke B, Methner A, Nykjaer A, Susens U, Fenger U, Herbarth B, Schaller HC (1998) Unique expression pattern of a novel mosaic receptor in the developing cerebral cortex. Mech Dev 70: 65–76.

Jacobsen L, Madsen P, Moestrup SK, Lund AH, Tommerup N, Nykjaer A, Sottrup-Jensen L, Gliemann J, Petersen CM (1996) Molecular characterization of a novel human hybrid-type receptor that binds the alpha2-macroglobulin receptor-associated protein. J Biol Chem 271: 31379–31383.

Jacobsen L, Madsen, P, Jacobsen C, Nielsen MS, Gliemann J, Petersen CM (2001) Activation and functional characterization of the mosaic receptor SorLA/LR11. J Biol Chem 276: 22788–22796.

Jacobsen L, Madsen P, Nielsen MS, Geraerts WP, Gliemann J, Smit AB, Petersen CM (2002) The sorLA cytoplasmic domain interacts with GGA1 and -2 and defines minimum requirements for GGA binding. FEBS Lett 511: 155–158.

Khvotchev M, Sudhof TC (2004) Proteolytic processing of amyloid-beta precursor protein by secretases does not require cell surface transport. J Biol Chem 279: 47101–47108.

Marcusson EG, Horazdovsky BF, Cereghino JL, Gharakhanian E, Emr SD (1994) The sorting receptor for yeast vacuolar carboxypeptidase Y is encoded by the VPS10 gene. Cell 77: 579–586.

Motoi Y, Aizawa T, Haga S, Nakamura S, Namba Y, Ikeda K (1999) Neuronal localization of a novel mosaic apolipoprotein E receptor, LR11, in rat and human brain. Brain Res 833: 209–215.

Nielsen MS, Madsen P, Christensen EI, Nykjaer A, Gliemann J, Kasper D, Pohlmann R, Petersen CM (2001) The sortilin cytoplasmic tail conveys Golgi-endosome transport and binds the VHS domain of the GGA2 sorting protein. Embo J 20: 2180–2190.

Nothwehr SF, Bruinsma P, Strawn LA (1999) Distinct domains within Vps35p mediate the retrieval of two different cargo proteins from the yeast prevacuolar/endosomal compartment. Mol Biol Cell 10: 875–890.

Nykjaer A, Willnow TE (2002) The low-density lipoprotein receptor gene family: a cellular Swiss army knife? Trends Cell Biol 12: 273–280.

Nykjaer A, Lee R, Teng KK, Jansen P, Madsen P, Nielsen MS, Jacobsen C, Kliemannel M, Schwarz E, Willnow TE, Hempstead BL, Petersen CM (2004) Sortilin is essential for proNGF-induced neuronal cell death. Nature 427: 843–848.

Offe K, Dodson SE, Shoemaker JT, Fritz JJ, Gearing M, Levey AI, Lah JJ (2006) The lipoprotein receptor LR11 regulates amyloid beta production and amyloid precursor protein traffic in endosomal compartments. J Neurosci 26: 1596–1603.

Peraus GC, Masters CL, Beyreuther K (1997) Late compartments of amyloid precursor protein transport in SY5Y cells are involved in beta-amyloid secretion. J Neurosci 17: 7714–7724.

Petersen CM, Nielsen MS, Nykjaer A, Jacobsen L, Tommerup N, Rasmussen HH, Roigaard H, Gliemann J, Madsen P, Moestrup SK (1997) Molecular identification of a novel candidate sorting receptor purified from human brain by receptor-associated protein affinity chromatography. J Biol Chem 272: 3599–3605.

Scherzer CR, Offe K, Gearing M, Rees HD, Fang G, Heilman CJ, Schaller C, Bujo H, Levey AI, Lah JJ (2004) Loss of apolipoprotein E receptor LR11 in Alzheimer disease. Arch Neurol 61: 1200–1205.

Schmidt V, Sporbert A, Rohe M, Reimer T, Rehm A, Andersen OM, Willnow TE (2007) SorLA/LR11 regulates processing of amyloid precursor protein via interaction with adaptors GGA and PACS-1. J Biol Chem 282: 32956–32964.

Seaman MN (2004) Cargo-selective endosomal sorting for retrieval to the Golgi requires retromer. J Cell Biol 165: 111–122.

Seaman MN (2005) Recycle your receptors with retromer. Trends Cell Biol 15: 68–75.

Small SA, Gandy S (2006) Sorting through the cell biology of Alzheimer's disease: intracellular pathways to pathogenesis. Neuron 52: 15–31.

Small SA, Kent K, Pierce A, Leung C, Kang MS, Okada H, Honig L, Vonsattel JP, Kim TW (2005) Model-guided microarray implicates the retromer complex in Alzheimer's disease. Ann Neurol 58: 909–919.

Spoelgen R, von Arnim CA, Thomas AV, Peltan ID, Koker M, Deng A, Irizarry MC, Andersen OM, Willnow TE, Hyman BT (2006) Interaction of the cytosolic domains of sorLA/LR11 with the amyloid precursor protein (APP) and beta-secretase beta-site APP-cleaving enzyme. J Neurosci 26: 418–428.

Taira K, Bujo H, Hirayama S, Yamazaki H, Kanaki T, Takahashi K, Ishii I, Miida T, Schneider WJ, Saito Y (2001) LR11, a mosaic LDL receptor family member, mediates the uptake of ApoE-rich lipoproteins in vitro. Arterioscler Thromb Vasc Biol 21: 1501–1506.

von Arnim CA, Tangredi MM, Peltan ID, Lee BM, Irizarry MC, Kinoshita A, Hyman BT (2004) Demonstration of BACE (beta-secretase) phosphorylation and its interaction with GGA1 in cells by fluorescence-lifetime imaging microscopy. J Cell Sci 117: 5437–5445.

Westergaard UB, Sorensen ES, Hermey G, Nielsen MS, Nykjaer A, Kirkegaard K, Jacobsen C, Gliemann J, Madsen P, Petersen CM (2004) Functional organization of the sortilin Vps10p domain. J Biol Chem 279: 50221–50229.

Xu H, Greengard P, Gandy S (1995) Regulated formation of Golgi secretory vesicles containing Alzheimer beta-amyloid precursor protein. J Biol Chem 270: 23243–23245.

Yamazaki T, Selkoe DJ, Koo EH (1995) Trafficking of cell surface beta-amyloid precursor protein: retrograde and transcytotic transport in cultured neurons. J Cell Biol 129: 431–442.

Yamazaki H, Bujo H, Kusunoki J, Seimiya K, Kanaki T, Morisaki N, Schneider WJ, Saito Y (1996) Elements of neural adhesion molecules and a yeast vacuolar protein sorting receptor are present in a novel mammalian low density lipoprotein receptor family member. J Biol Chem 271: 24761–24768.

Zhu Y, Doray B, Poussu A, Lehto VP, Kornfeld S (2001) Binding of GGA2 to the lysosomal enzyme sorting motif of the mannose 6-phosphate receptor. Science 292: 1716–1718.

Zhu Y, Bujo H, Yamazaki H, Ohwaki K, Jiang M, Hirayama S, Kanaki T, Shibasaki M, Takahashi K, Schneider WJ, Saito Y (2004) LR11, an LDL receptor gene family member, is a novel regulator of smooth muscle cell migration. Circ Res 94: 752–758.

Index

LIST OF PREVIOUSLY PUBLISHED VOLUMES IN THE SERIES RESEARCH AND PERSPECTIVES IN ALZHEIMER'S DISEASE

A. Pouplard-Barthelaix et al. (Eds.) (1988) *Immunology and Alzheimer's Disease*
P.M. Sinet et al. (Eds.) (1988) *Genetics and Alzheimer's Disease*
F. Gage et al. (Eds.) (1989) *Neuronal Grafting and Alzheimer's Disease*
F. Boller et al. (Eds.) (1989) *Biological Markers of Alzheimer's Disease*
S.I. Rapoport et al. (Eds.) (1990) *Imaging, Cerebral Topography and Alzheimer's Disease*
F. Hefti et al. (Eds.) (1991) *Growth Factors and Alzheimer's Disease*
Y. Christen and P.S. Churchland (Eds.) (1992) *Neurophilosophy and Alzheimer's Disease*
F. Boller et al. (Eds.) (1992) *Heterogeneity of Alzheimer's Disease*
C.L. Masters et al. (Eds.) (1994) *Amyloid Protein Precursor in Development, Aging and Alzheimer's Disease*
K.S. Kosik et al. (Eds.) (1995) *Alzheimer's Disease: Lessons from Cell Biology*
A.D. Roses et al. (Eds.) (1996) *Apolipoprotein E and Alzheimer's Disease*
B.T. Hyman et al. (Eds.) (1997) *Connections, Cognition and Alzheimer's Disease*
S.G. Younkin et al. (Eds.) (1998) *Presenilins and Alzheimer's Disease*
R. Mayeux and Y. Christen (Eds.) (1999) *Epidemiology of Alzheimer's Disease: From Gene to Prevention*
V.M.-Y. Lee et al. (Eds.) (2000) *Fatal Attractions: Protein Aggregates in Neurodegenerative Disorders*
K. Beyreuther et al. (Eds.) (2001) *Neurodegenerative Disorders: Loss of Function Through Gain of Function*
A. Israël et al. (Eds.) (2002) *Notch from Neurodevelopment to Neurodegeneration: Keeping the Fate*
D.J. Selkoe, Y. Christen (Eds.) (2003) *Immunization Against Alzheimer's Disease and Other Neurodegenerative Disorders*
B. Hyman et al. (Eds.) (2004) *The Living Brain and Alzheimer's Disease*
J. Cummings et al. (Eds.) (2005) *Genotype – Proteotype – Phenotype relationships in Neurodegenerative Disease*
M. Jucker et al. (Eds.) (2006) *Alzheimer: 100 Years and Beyond*
D.J. Selkoe et al. (Eds.) (2008) *Synaptic Plasticity and the Mechanism of Alzheimer's Disease*

Printing: Krips bv, Meppel, The Netherlands
Binding: Stürtz, Würzburg, Germany